In Command of Guardians:
Leadership for the Commun... ....

Eric J. Russell

In Command of Guardians: Executive Servant Leadership for the Community of Responders

Second Edition

 Springer

Eric J. Russell
Department of Emergency Services
Utah Valley University
Orem, UT, USA

ISBN 978-3-030-12492-2   ISBN 978-3-030-12493-9   (eBook)
https://doi.org/10.1007/978-3-030-12493-9

This Springer imprint is published by the registered company Springer Nature Switzerland AG
The registered company address is: Gewerbestrasse 11, 6330 Cham, Switzerland

For Pop
James F. Russell
11/13/1944–9/7/2016
As a child, I never had to look for a hero
As a son, I never questioned if I was loved
As a man, I never had to seek a mentor
As his witness, I never can say that granite
doesn't crack

A friend asked me years ago, "Besides your
Dad, who do you look up to?"
My reply, "No one."

Foreword

The emergency services have a long history of being at the ready and fulfilling their missions by responding to crises of all sizes and magnitude. Today, the emergency services have developed into quite a diverse vocation ranging from volunteer responders who serve their communities with fidelity and courage to those who are highly trained professionals whose careers are emergency prevention, response, enforcement, and recovery. The emergency services have enjoyed a tremendous amount of development in the applications of science and technology in "the business." However, the focus on science, technology, engineering, and mathematics (STEM) has resulted in some serious side effects for both the responders and the citizens they serve.

As society at large moves at an ever-more fast pace and grows in technological complexity, the emergency services begin to require a different kind of leadership. Emergency services agencies have become complex organizations that have historically run on a paramilitary model of management, adopted from colonial times, and kept to muster, organize, and direct the responders on the chaotic scenes to which they were expected to renormalize. In early times, fire departments were mere assemblies of local men who responded when an alarm was made. Now, they have evolved into personnel and equipment that require many layers of responsibility and multimillion-dollar budgets. As a result, the focus for governments is to find those among the ranks who can manage the personnel, logistics, and budgets that provide services to the whole community. That is where, like business, the acquisition of technology in terms of computer record-keeping and communications becomes necessary to manage the affairs of the department within the context of the whole municipal organizational structure.

Governments are bureaucracies for the sole purpose of centralizing power, accountability, and maintaining control of public assets. For example, the Declaration of Independence, the foundation of America, expresses clearly that separation was needed due to the oppressions of the British Monarchy—skepticism, and even sometimes cynicism, about government is part of the fabric of American culture. In fact, it almost seems to make people feel "more American" when they speak about "the government" in derogatory ways. Bureaucracies are inefficient, at times nonsensical, and the citizen typically interacts with the bottom ranks that have little or no power to do anything outside of the prescribed policies and protocols. Unique problems require unique solutions, and low-level bureaucrats are not empowered to innovate in their roles. But because bureaucracies are hierarchical in structure (and, I would argue, dominated by patriarchal perspectives, attitudes, and functioning), most of the messiness of community problems needs effective evaluation, equitable prioritization, and solutions that serve the citizens rather than the rote public policy book. Governments need both sufficient strength and flexibility in order to withstand the demands of an ever-developing and complex society.

Governments need to be resilient in order to serve the citizens under their care. The aforementioned ambivalent relationship citizens have with "the government" whitewashes over the fact that many societies are very decentralized in their governmental structures, particularly on the local level. But that does not change the mental models of the citizens whose most dominating lessons in civics come from major news outlets. Again, technology has enabled the news to not only report what is going on in the society but also shape society in its perspectives, attitudes, and values. As such, when a citizen walks into his or her local government center with a problem, bill to pay, or concern, he or she is met oftentimes with an objectifying experience, not simply because the low-level bureaucrat with whom he or she meets is an oppressor but rather that preconceived expectations are loaded into the hearts and minds of the citizens. Moreover, the pat answers of "that is not allowed by policy" or "you're in the wrong line, you'll need to move to that one" infuriate the citizen who has little time to serve the government rather than be served by it. The clerk does not enjoy this kind of interaction either; he or she would love to enjoy just a little bit of latitude to adapt and overcome some of the rigidity that is inherent in bureaucratic structures and processes.

The emergency services exist within the governmental bureaucratic structure but add to it the paramilitary styles that were found useful in managing emergency scenes. Rank structures are intended to provide clear lines of authority and levels of responsibility. Emergencies are those human events that are time-pressured and will likely result in several harmful consequences if conditions are not adequately mitigated in time. On the scene of a crisis or disaster, incident management system structures and functions have improved the efficiency and effectiveness of emergency response. The command and control style of leadership works well, and everyone understands that time constraints and threats of human life leave little room for deliberation or debate about incident operations plan. Therefore, it is vital to have a management team whose knowledge of emergency work and management systems produces the best outcomes that each situation will allow. Competent authoritative leaders are needed to serve in these roles.

Diversity and inclusion in the emergency services is a slow-going process. Even with the advent and development of advanced technology and training, the "big man" still seems to be the idealized responder in the mental models of the organizations and the communities they serve. But all things continue to evolve and emergency services are changing. Patriarchies position power centralized at the top levels that often encourage cronyism—a practice that tends to foster toxic leadership.

Those in power tend to self-insulate by surrounding themselves with like-minded people.

Changes always happen with a certain amount of energy, tension, and pressure between forces. Traditions are used to stabilize organizations and groups so that their evolution is manageable and sensible. One thing that we know from trauma studies is that one's restored sense of stability and control is a vital aspect of recovery from a critical incident. Innovations are called for when the traditional, customary, or established doctrines and rituals no longer suffice. The more conventional the organization, the more discomfort and direct dysfunction from obsolete policies and procedures must be endured before the necessity of innovation becomes inevitable. As such, the innovators and early adopters of new ideas seem to come up against formidable resistance from the conventional-minded when trying to make changes in the organization.

For one to be a citizen of a nation is a sacred thing. For one to be a member of a community is wholesome and cultivating. Our national ethos was once about citizenship but has been steered toward consumerism. Advertising jingles as old as "hold the pickles, hold the lettuce, special orders don't upset us... have it your way" have enchanted the citizen because the "customer is always right." Certainly, business figured out that it could better serve its customers through being flexible in its production, sales, and delivery of its commodities. Public services needed a boost in public relations but not when this is merely a reification of propaganda and the manufacturing of consent. Rather, the government as a whole needs a new relationship with its citizens. The anti-governmental sentiments that the founders of America expressed and promoted were based on aristocratic command and control. Undoubtedly, Benjamin Franklin was a man of liberty and innovation. So too, emergency responders must be prepared, supported, and empowered to serve citizens. Being a taxpayer does not make one a customer of the government any more than does being a citizen of the USA endow him or her with the inalienable rights of a human being. Public servants must come into their roles with a desire to serve rather than a desire for power and control, even when power and control to renormalize a situation are needed.

In the current state of affairs, the dominant bureaucratic ethos advanced with technocratic speed and has made government more efficient in many ways but less and less personal for the citizens and those who serve them. Bureaucracy still functions to ensure stability, accountability, and control of public assets. Technology allows for many systems of management and communication to move faster than ever before. But, we cannot lose sight of the fact that the emergency services are part of the helping professions. We cannot lose sight of the fact that emergency services are about serving people. We cannot lose sight of the reality that emergency responders are citizens and come from among the citizenry to serve their communities.

Some of the innovations now needed in the emergency services must come from some other disciplines. We need some new ideas and perspectives to improve the lives of our responders to make them more resilient and personally and professionally efficacious. Seemingly, the best leaders for the emergency services are ones that come into leadership with a desire to serve. Leadership must cultivate, care, and empower its people so that services rendered can best aid our citizens. With a priority on caring for responders first, the responders will deliver the services efficiently, efficaciously, and with equity to the community.

We must build and sustain a better emergency responder for today. It is not just about making them work faster, harder, and longer than ever before through management systems and applied technology. Rather, it is about re-humanizing them in the context of their missions. So often, people do not even look past the uniforms. It takes a photograph in a magazine at the scene of a terrorist event or natural disaster, a grief-stricken responder holding an injured or dead child, to remind us all of their humanity and of ours. The traditional ethos of the emergency services tells us that these are the narrow and appropriate times to see our tough responders cry. But there needs to be more.

The latest literature on stress and resilience in the emergency services indicates that the leading source of stress for responders is organizational hassles—bureaucratic mechanisms. The objectifying process is said to offer fair, equitable, and accountable management of agency functions. All too often, however, the wheels of a bureaucratic process roll over the top of responders. Instead of doing the right thing, higher-ups often "do things right" within the parameters of the policy and procedure manual but are (in the very least perceived to be) executing personal agendas of power and control. This erodes social support and often fosters toxic competition for promotion, special assignment, and status among the ranks. The ethos of command and control drowns the desire to serve, and responders can, and will, lose sight of their mission. We do not hire apathetic, cynical, and depressed responders—we create them.

What emergency workers know is that they desire to become a firefighter, paramedic, emergency medical technician, police officer, etc. Like other caring professions, the knowledge, skills, and abilities are practiced in a way that often transforms the citizen into the rescuer, protector, and caregiver. Responders know that regardless of how vast and full their preparatory training is, they become the firefighter by fighting fires, the police officer by enforcing the law, the paramedic by providing prehospital care, and the hazardous materials or advanced rescue technician through engaging hazardous environments. For leaders to be servant leaders, they need to practice it. Like anything else, it does not come all at once, and many executive officers, administrators, and mid-level supervisors may be skeptical and hesitant about resurrecting their desire to serve and live it out bodily. It takes a lot of courage to set aside command and control in the office and replace it with service. All of the science, technology, engineering, and mathematics in the world cannot replace the values of care, compassion, and empowerment of human beings.

Dr. Eric Russell began his journey decades ago serving as a fire and rescue specialist for the US Air Force (USAF). His start did not begin in the usual paramilitary environment that most community firefighters serve in but rather in the military. He learned to understand and promotes the values of a command and control structure in the context of combat, emergency, crisis, disaster, etc. However, he experienced some of the by-products of the patriarchy that offended his sensibilities as one who came into the work with a desire to serve. Dr. Russell is one who describes himself as an innovator balanced with a reverence for tradition. He recognizes the value of stability while acknowledging that life is an ever-changing dynamic process. That is, evolution is a process that requires a certain degree of energy, intellect, and fortitude to get static structures and systems to adapt to changing environmental conditions.

Dr. Russell is an educator, researcher, and scholar in the emergency services. Not content with the current state of the command and control ethos, he sought to become an expert in servant leadership so that its application could benefit the emergency services. But the lessons Dr. Russell learned and the paradigmatic shift he experienced have actually led him to see that his initial vision could be broader and deeper than he imagined at first blush. In the continual dialogue about increasing the professionalism and efficacy of emergency response, Dr. Russell's work actually lends itself to initiating sustainable organizational change. Changes that can occur at all levels of emergency work but also transform the relationship that the public has with its responders.

In the past, various leaders in the emergency services paved the way for people like Eric to go to college and advance their education to accompany their work experience. Many leaders secured business degrees and began importing the ethos of business with concepts such as customer service, performance-based evaluation, risk management, human resources, zero-balance budgeting, and more. While well-intended, the principles of business came in the form of spreadsheets, quantified measures for everything, and data-driven decision-making. Now, this all sounds good and certainly works to some degree in business, but the objective of business is not just to protect the bottom line but also to profit and grow it. Public service is about serving citizens, not customers.

Salt Lake City, UT, USA   Rodger E. Broomé

Preface

Over the past decade, empirical works discussing servant leadership for the emergency services have been published; however, they primarily focus on the responder. This book advances these works, focusing now on how the practice of servant leadership actually works within the emergency services and the benefits derived from it. Though this work can be used at all levels, it is specific to command level leaders within the emergency services career field.

To bring the power of this philosophical approach toward leadership to life, this work grounds itself in the foundational aspects of identified servant leadership characteristics (Spears 2010), virtuous constructs (Patterson 2003), and attributes (Laub 1999). The text demonstrates how the many aspects of the philosophy come together as a system in order to strengthen the community of responders. Additionally, this work attempts to make the case to industry and academia with regard to increasing servant leadership-specific curriculum for current and aspiring emergency services officers.

The second edition expands *In Command of Guardians: Executive Servant Leadership for the Community of Responders* by infusing research findings into relating chapters throughout the text in order to bring to life the philosophy. This is done to merge current research and theoretical ideas involving responders and servant leadership.

Aspects of a qualitative research study involving the interpretation of company-level fire and emergency services officers regarding the role and characteristics of leadership are used in multiple chapters of the text allowing for a richer understanding as to why the philosophy matters. Russell et al. (2015) originally published the research findings in the journal *Servant Leadership: Theory & Practice* which are used in areas of this book with permission of the journal's editor. Russell et al.'s (2015) grounded theory study involved 15 uniformed and sworn fire and emergency services officers as participants. An overview of the study is given in Chap. 4 to spotlight the background of the research as well as a short discussion of the findings as they relate to servant leadership literature. Because the nature and the impact of this research project enhance the heart of this book, the author is thankful for the permission to reuse the work.

In this second edition, new chapters have been added to discuss the concepts of the responder's cycle of trust and fostering responder servant-followership: concepts in need of cultivation throughout the professions. Each chapter goes in depth on the different aspects as they relate to executive-level leadership, how servant leadership impacts responders, and the successes realized by those who chose to be servant leaders.

Additionally, chapter summaries offer case studies and questions as meditative exercises, giving the reader a rich secondary learning experience that comes from the reflective critical-thinking process. The second edition of *In Command of Guardians: Executive Servant Leadership for the Community of Responders* grounds itself in existing empirical works involving both servant leadership and its place within the emergency services profession. It is the goal of the author to present the philosophy in such a way so the reader not only gleans a rich understanding and applicability of the concepts but also is inspired to foster servant leadership within their own organizations.

Orem, UT, USA   Eric J. Russell

References

Laub, J. (1999). Assessing the servant organization: Development of the servant organizational leadership assessment (SOLA) instrument. (Doctoral dissertation). Retrieved from ProQuest Dissertation and Theses Database. (UMI No. 9921922).

Patterson, K. (2003). Servant leadership: A theoretical model (Doctoral Dissertation). Available from ProQuest Dissertation and Theses Database. (UMI No. 3082719).

Russell, E., Broomé, R., & Prince, R. (2015). Discovering the servant in fire and emergency services leaders. *Servant Leadership: Theory & Practice, 2*(2), 57–75.

Spears, L. (2010). Servant leadership and Robert K. Greenleaf's legacy. In K. Patterson & D. van Dierendonck (Eds.), *Servant leadership: Developments in theory and research* (pp. 11–24). New York: Palgrave Macmillan.

Acknowledgments

I would first like to thank my publisher Springer Nature, especially my editor Paul Drougas who saw the need to expand the work and develop a second edition of *In Command of Guardians: Executive Servant Leadership for the Community of Responders*. In addition, I would like to thank my personal line editor, Abbi McKall Mills, for making the work beautiful.

Thanks to Emily Hough, Editor-in-Chief of the *Crisis Response Journal*, and Dr. Phillip Bryant, Executive Editor, *Servant Leadership: Theory & Practice*, for allowing my previous works to become the foundations on which this work was built.

Thanks to Dr. Rodger E. Broomé for being a friend, a co-researcher, and a mentor. Our dialogue and ideas inspire works like this; it comes from a shared desire to strengthen the emergency services and a love for those who serve.

Thanks to the innovators and leaders of servant leadership, especially Pat Falotico, Dr. Larry Spears, Dr. Kathleen Patterson, Dr. Don Frick, Dr. Brice Winston, Dr. Dirk van Dierendonck, Duane Trammell, Dr. Ken Blanchard, and Dr. Jim Laub, as well as other teachers, researchers, and authors that work tirelessly to dig deeper into Greenleaf's philosophy.

Thanks to Dr. R. Jeffrey Maxfield and Dr. Jamie L. Russell, for helping me expand this second edition of *In Command of Guardians: Executive Servant Leadership for the Community of Responders*. I am grateful for their help and expertise.

Thank you to the guardians and those in command; this work is for you.

Thanks to my family, especially my mother, for always supporting my work and believing in me.

To my wife, my happiness, and soul mate. You encourage and strengthen me to always strive to be better than I was yesterday. You are my gift, the love of my life.

To God goes all the Glory.

Contents

About the Author

Eric J. Russell is an Associate Professor with Utah Valley University's Department of Emergency Services. He retired early from the Department of Defense, Air Force Fire and Emergency Services, as a Captain with a combined service of active duty military and Department of Defense. He has a Doctorate of Education in Organizational Leadership and a Master of Science in Executive Fire Service Leadership from Grand Canyon University. In addition, Eric is also a Certified Homeland Protection Professional (CHPP) from the National Sheriffs' Association and the Global Society of Homeland and National Security Professionals (GSHNSP) and was awarded a Graduate Certificate in Homeland Security Studies from Michigan State University.

Eric and his wife Jamie live in Salt Lake City, UT.

Chapter 1
In Command of Guardians

*Great things are not done by impulse, but by a series of small
things brought together.*

—*Vincent van Gogh*

Abstract This chapter introduces the philosophy of servant leadership to those responsible for a community of emergency responders. This introduction spotlights the potential that servant leadership can have for improving the lives of guardians. It begins by offering a pathway for self-reflection regarding the emergency services leader's desire to serve—a desire which brought them to the career field years before. It defines the community of responders as a quasi-family-like paramilitary world in which responders live and operate. Serving this community, as well as on-scene command-and-control, are the greatest responsibilities facing the emergency services leader. The career of a first responder is one of noble sacrifice and service; servant leadership is offered as a way to honor that service.

Where you are at today didn't happen by mistake. Your position and authority formed over time. Who you are today stems from a sequence of events and experiences. This journey didn't begin with you yearning to be a leader—it began with your desire to serve (Russell 2014a). Long ago, before it was even possible to imagine what the profession entailed, you simply desired a life of serving others in their time of need. The money didn't matter. For many, even the specific organization didn't matter. All you longed for was an opportunity to be a part of something greater then self. You wanted to save the day.

Take a moment and think back to your time in the academy, when this was all new to you. Think about your technical courses, and your academic studies. You looked at those that did the work with awe. You hung on every word uttered by your instructors, professors, and preceptors. You imagined serving with them. You longed

1
E. J. Russell, *In Command of Guardians: Executive Servant Leadership for the Community of Responders*, https://doi.org/10.1007/978-3-030-12493-9_1

to hear the "there I was" war stories. This is where the journey began and why you chose this path. It started not at the top of the ladder, but rather, at the very bottom.

Today you find yourself in command. You are a leader. You are the one they look to. Now your followers hang on your words; they desire to hear your "there I was" tales. They long for your mentoring. They want your approval. They want your protection. You are now in command of guardians.

To be in command of guardians means one has been gifted the responsibility, and therefore power, over those who willingly serve and sacrifice for strangers. Regardless of whether you achieved this position by election, promotion, or appointment, you are accountable for their wellbeing. Every decision you make will affect them, be it the closing of stations, shifts in staffing, implementation of new policies, or changes to operations. Each decision you make impacts the community of responders. This is a group built upon traditions, holding tight to the familiar and finding comfort in the known. So much of this profession requires problem solving in the face of the unknown. Problems arise in this profession when there's nothing in need of solving, when something cannot be solved, or when sudden changes, perhaps unexpected, occur.

An emergency services leader must not only realize, but also, help his or her followers recognize that the average person can never see the world from the perspective of the responder. Hence the reason that you as a leader must be a steward for them—not just within the community of responders, but also, to the public itself. In his work *The Ethic of Strength*, Greenleaf (1996) wrote,

> Very few people accept that this is a dangerous world [morally, physically, intellectually] and hence they do not choose to be aware of where they are, who they are, what kind of world they live in, or what its traps and hazards are (p. 38).

Though at the time he wasn't writing specifically to the emergency services, his words identify an intractable barrier existing between emergency responders and the citizenry. Additionally, it highlights the stark differences needed to lead a community of responders.

For example, because of their experiences, the responder encounters flight-or-flight before the civilian. What's more, the responder will make certain life decisions based upon their previous experiences. The public is predominantly ignorant to the fact that their story will become a responder's story; their decisions in life can have a negative impact on the responder. Case in point, the driver that chooses to not wear a seatbelt or a motorcyclist that chooses to not wear a helmet. Though it is the individual's decision, the consequences when they are involved in accidents become gruesome realities responders must navigate.

To go one step further with this topic, think about just how difficult it is for you and for many other operators to sit with one's back towards the door. The responder is keenly aware of their surroundings and the people around them because of an understanding of human nature and just how quickly things can change. Within minutes of being a patron of a business or a guest in someone's home, responders seem to know the location of egress points as well as potential hazards and threats. It must never be forgotten that unlike a knot, experiences cannot be undone.

So, what does it mean to be in command of guardians? It begins with an understanding of the structure of the emergency services profession. The nature of emergency services work involves long periods of routine monotony interrupted by moments of traumatic, adrenaline-pumping chaos. This is the world the professional responder must navigate and you must lead. It is a balancing act between paramilitary governmental structures with patriarchal ranks coupled with family-like living arrangements. Between calls, responders live in close-knit sibling-like relationships with clear demarcations between the quasi-family unit and a hierarchical command system. Leaders will find themselves giving life-or-death orders to crews and teams on an emergency call that interrupted their dinner together. This is a comparable reality for emergency services personnel across the globe; it is the collective familiar world that they live in; it is what is known as the community of responders.

To be in command of guardians is one of the two greatest obligations facing leaders within the emergency services. The other is the management of emergency scene command and control. The profession succeeds at preparing managers for the role of incident command through rigorous academic and training programs coupled with vast amounts of continuous education. However, when it comes to areas involving the preparation of individuals for non-emergency scene leadership within the emergency professions, research reveals a need for improvement (Russell 2014a). There are some theories as to why this area is neglected, such as a concentration on managerial functions within emergency services officer curriculum, or the assumption that one's rank automatically makes them a capable leader. Such concentrations and assumptions create vulnerabilities for the profession's most valuable asset: its people.

This lack of focus on the role of leadership within the emergency services led to the development of this work. Specifically, this book spotlights the reasons why leaders need to focus more on this community of responders and with it, the hope of motivating those same leaders to take a more active role in honing servant leadership skills that benefit the needs of people. So why servant leadership and not another leadership theory? It is because servant leadership characteristics, constructs, and attributes mimic those found in most emergency services responders (Russell 2014a). Research has shown that the characteristics, constructs, and attributes forming the philosophy of servant leadership are the same ones that brought the responder to this profession (Russell 2014a, b; Russell et al. 2015). They just need to be called out, revived, and brought to the forefront.

Currently, existing empirical works present the relationship between servant leadership philosophy and the emergency services; however, they don't explain how being a servant leader improves the lives of emergency responders. *In Command of Guardians* shows the reader how the qualities associated with the philosophy serves the community of responders. Naturally cultivating servant leadership qualities positively influences one's ability to foster the community. This relates back to what Greenleaf (1970) called your natural desire to serve.

Defining the Guardian

Throughout human history there have been those willing to step forward to shield the group (Hart and Sussman 2005). Brave persons called to a role to protect and serve others. It began in tribes where certain individuals became the "watchman" standing guard on the hill (Hart and Sussman 2005). These individuals transcended the average citizenry to become the guardians of the people, at times willing to sacrifice self for the community. The guardians have always been there—they stand outside the lines and respond in times of need (Hart and Sussman 2005).

Today, the guardians are those who take the oath of service as professional emergency responders, commonly referred to as firefighters, police officers, emergency medical technicians, and rescue personnel. These are the modern guardians, ordinary individuals with a desire to serve, willing to step forward to serve others in extraordinary times of need. The guardian role has evolved over time and will continue to evolve; however, what remains is the desire of the few to protect and defend the many.

Leading the Alpha

The emergency services profession attracts individuals that are commonly described as alphas. They are usually mentally strong, courageous, and physically fit individuals. Thus, there is a culture that exists within the community of responders that is inherently tough and prideful in its self-reliance. As an emergency services leader, you came up through this culture. It is a part of who you are. It is possible that if others were asked to define you, they most likely would describe you as an alpha, possibly an apex-alpha.

Peter Drucker once said, "Culture eats strategy for breakfast." It is not the desire of this book to change or remove this alpha culture. Such strength and resilience are needed to perform the work. Emergency service professions are not good career choices for the timid and the meek. Instead, what this work sets forth to do is put that alpha personality into perspective as it relates to servant leadership. Moreover, to show that there is a difference between what it means to be an alpha and its often confused and destructive cousin "hyper-masculine", a state of being where emotions, feelings, and vulnerability are seen as weaknesses (Kimmel 2011).

Leadership must acknowledge that the alpha is still human and therefore vulnerable to psychological traumas and mental stresses that come from the work (Koran 2016). Though you can train and educate guardians and warriors to be resilient, you cannot train away all vulnerability (Bartone et al. 2008). To bring this point home, one needs to look no further than the elite United States Special Operations. Society seems to want to believe that when it comes to alphas, that the best-of-the-best rises above the impact of psychological trauma (Shanker and Oppel 2014). However, such a belief is a myth. Research has shown that even top warriors are prone to post-traumatic stress disorder, psychological issues, and substance abuse (Hing et al. 2012; Kilgore et al. 2008; Skipper et al. 2014).

Defining the Community of Responders

The community of responders originates spontaneously from a culture of belonging and a shared desire to serve. As stated earlier, fostering this community is one of the greatest responsibilities facing the emergency services leader. The intention being that the community becomes a healthy place that strengthens individual responders, allowing for organic growth-providing interventions long before responders ever face their first traumatic experience. In doing so, such community practices establish a proactive, rather than a reactive, approach. This community must be people-centered with a focus on strengthening responders' sense of belonging, mental resiliency, ability to heal and cope, as well as personal and professional growth.

To understand the importance of fostering such a community one needs to look no further than the statistics published by the United States National Fallen Firefighter Foundation. The foundation estimates that professional responders are three times more likely to take their own life than to be killed in a line-of-duty incident. This reality leaves one asking why. A possibility is that it begins with a need for bureaucracies to find quantifiable solutions to qualitative problems, a simple algorithm of "if this, do that". There isn't malice behind this idea, just overburdened systems relying overwhelmingly on the medical model of treating symptoms with the hope of curing a responder's pain.

More often than not, governmental agencies intervene with well-intentioned, post-incident, bureaucratic-laden, critical incident stress debriefings and mental health services for responders. However, according to an interview with emergency services psychologist Dr. Rodger E. Broomé, if a system is intervening after the fact then it is not effective. Dr. Broomé notes that although these services are important to an overall mental health program, it is much more vital to strengthen a responder's pre-incident post-traumatic growth, creating resiliency within the emergency response community before the incident occurs. This involves nurturing a community of responders in a way that promotes physical, mental, and spiritual health: the cornerstones of mental resiliency, healing, and growth. It is here where servant leadership can have the greatest impact due to its focus on the needs of the individual.

Simply based on human history and natural occurrences, we know that tragedies are not going away. In fact, we know there will always be unavoidable emergency scenarios that will require responses. However, according to Kirschman (2004, 2006) the issue is not necessarily a traumatic scene. Kirschman (2004, 2006) finds that it is misguided to focus on the inevitable work experiences as the cause of responders' psychological issues. The fact is, most professional responders do this work because they possess a desire to serve, and want to respond and render aid. That's why Kirschman (2004, 2006) argues that it is not about the calls, but rather, navigating toxic bureaucracies that lead to unhealthy work environments. These are the same systems that exacerbate post-traumatic stress disorders in combat veterans by trying to normalize abnormal circumstances after returning home from combat.

Responders, like veterans, find themselves having to navigate the social construct of "normal" within society. They are left questioning if they themselves are

normal after their experience. The truth is, there is no normal, just the human created "Norman Rockwell" normal-fallacy that responders are awoken to once they
navigate human tragedy and loss. When we wield our realities into that which we
desire, we convince ourselves to see things not as they are, but rather, how we desire
them to be; i.e. normal. This can be a pitfall to the responder's mental health when
this social construct plays a part in convincing them that how they are feeling is
wrong, leaving them unable to see it is quite the opposite. We want to pretend that
it is not normal and that they need to snap-out-of-it, all the while forgetting that the
entire concept of normal is absurd. The philosophy of servant leadership is a way to
overcome these toxic pitfalls by serving the community of responders' needs as
they are, not as others think they should be.

Traumatic experiences within the emergency services are inevitable; a leader
does not have the power to change that. Furthermore, responders enter this career
knowing full well the need to operate in these conditions. Though it isn't possible
for newbies to truly understand the nature of real-world emergency response expo-
sure, thanks to technology and war-stories, most have at least an inclination of what
the work entails. That being said, what isn't inevitable and can be changed within
bureaucracies are issues such as staffing needs instead of what responders really
need—some time off, a sincere listener, a system of people serving people not policies, physical-fitness activities trumping non-essential tasks, recognizing if an environment views feelings and emotions as weakness (or worse a mental illness in need
of treatment). These are just a few examples that support this argument as to why
emergency service leaders need to cultivate a community of responders in an
attempt to implement positive environments and to remove or at least reduce bureaucratic stumbling blocks.

This begins with fostering open dialogue, healthy time off, education and promotion of healthy living, on-duty physical fitness programs, and a review and replacement of policies that are not people-centric. As a result, responders will begin to
experience a community that says it is fine to not be fine, that it is a personal, and
expected, reaction to an abnormal experience. Additionally, a community where
responders know that it's also acceptable to not feel anything at all, that it is not
abnormal to feel nothing, but rather, abnormal and mentally unhealthy to try to feel
something one does not. Either way, responders need this community of health,
safety, and healing to become the place in which they land between incidents, a
community that is people-centric and focused on belonging. A community cultivated by servant leaders.

One factor that can inspire emergency services leaders to absorb the material in
this book, and afterwards foster such a community, is to become aware that tragedies and losses pile on. Over time, they weigh down the responder's psychological
backpack. Statistically, a victim of a tragedy will only undergo that experience once
in a lifetime. On the other hand, responders will experience new tragedies over-and-
over again in situations involving new victims.

Think of this concept in the following metaphor. One can be ordered to climb a
mountain wearing an 80lbs pack. This in itself is a difficult yet not impossible feat.
The person beginning the climb and struggling with both the grade and weight is the

emergency services responder. The one that ordered the person to climb decides to add to the pack, one pound at a time; this weight is a combination of the bureaucracy and the tragedies. The responder can accomplish the task and navigate the mountain because they are strong enough to carry the load; however, as the weight increases in the pack one pound at a time, they will eventually succumb to defeat and fail.

Governments spend a great deal of time and resources developing professional emergency services responders: individuals that physically and psychologically transcend the identity of the average citizenry to become that guardian on the hill, the watcher at the ready. This identity is vital to emergency response because without that transcendence an individual will break under the stress. However, this reality lends to the emergency responder the responsibility to mentally navigate devastating human events. Leadership must recognize that nothing in the emergency responders training and education prepares them for that secondary post-incident reflection of mankind at its worse.

There is not a program or school in existence that can remove the responder's humanity. No matter how well we train and educate them to become professional operators, they remain vulnerable human beings. Until we find a way to genetically engineer unfeeling robot-like responders, this reality will continue to exist. The very idea of removing a person's humanity means losing what it means to be human; to feel, to grieve, and to see our own mortality and that of loved ones in the lives of others. This is why fostering a community of responders is so vital: it becomes a place where it is okay to be human, a place where the responder can let down their public armor. A place where their needs are served so they in turn can serve the needs of others.

As we evolve as a society, our thinking changes. This includes the way we see tragedy and loss. Over time, human beings' recognition of right and wrong within advanced societies, coupled with an understating of compassion and humanity, culminates in a change of psyche. World War I was a great example of this. The horrors of that battle brought forth the concept of "shell shock". Today, the vast majority of people are repulsed by inhumane and brutal actions of individuals, groups, and governments including war and bloodshed. One way to understand this is the use of public executions when carrying out punishments of capital crimes. For millennia, societies have performed public executions of the convicted—some still do. However, advanced societies of today no longer carry out these events in view of the citizenry. Why? Sympathy for the convicted and the reality of what the punishment really is. Governments understand that regardless of how heinous the crime committed by the convicted, stomaching the taking of a life in an execution is simply too difficult for most. Even though some developed nations still use the death penalty as a form of punishment, the act is done in a private and sanitized manner. It is believed that if a public execution were to be carried out today, most citizenry would demand an end to the practice. Moreover, as it is with executions, the media sanitizes and censors what the public sees during tragedies, editing out the reality of death, deformity, and gore. Imagine for a moment the outcry if the public was privy to the violent impact of a bullet or a bomb.

Thanks to technology and a 24 h news cycle, the media will put forth images of emergency response personnel heroically responding and operating on the scenes of others' misfortunes. Regardless of where the incident occurs—a theater in Paris, a hotel in Mumbai, the coastline of Tōhoku, a nightclub in Orlando, a marketplace in Bagdad, a peaceful protest in Dallas, a concert in Las Vegas, or an elementary school in Sandy Hook— professional responders are called upon to mitigate the tragedy, putting the world of a stranger back together. They are not witnesses to the chaos, but rather, active participants. The tragedy of others becomes their tragedies. The stories of others become their stories. The experiences add to their psychological backpack. And this reality calls for leaders that are willing to help them carry its load.

Responders think and function under time-pressure and consequence, they hone their desire to serve through years of rigorous training and education, becoming highly skilled professionals. Yet despite all the training, desire, ability, and courage, the alpha never becomes less human. This is why they need a community of responders: it is the only way for individuals to overcome abnormal experiences, so they can return mentally, physically and spiritually healthy to the fight. The opportunity to lead this community of responders is a thing of trust— something a leader must desire to possess. That's what this book is about: how you as a leader can serve this community of responders.

Why Service Leadership

Responders are mentally and physically strong, well trained, and desire to do this work, so the looming question remains, why does the bureaucracy have such an affect on them (Volanti 2006)? The fact is that responders are just like any other person: vulnerable (Shakespeare-Finch 2006). Stated previously, it is not the work that causes the vulnerabilities, it is the responder's ability to cope and navigate the system (Shakespeare-Finch 2006). Again, the work is inevitable; bad things happen. But the system itself can change. This is possible with servant leadership, because it has been continuously proven to overcome toxic bureaucracies (Greenleaf 1977/2002). When servant leaders work to change the system, the coping skills of the individual responders can grow (Castellano and Everly 2006). As the system changes, followers are free to self-actualize as responders, growing mentally, physically, and spiritually stronger. In turn they can serve the needs of the public.

Responders deserve leaders that serve their needs, not from a position of servitude, but rather, a position of legitimate power. That is who the servant leader is: one who serves the needs of their followers from a position of legitimate power. The servant leader doesn't relinquish this position nor does serving their followers reduce their status (Hunter 2004). The servant leader understands that when a follower's needs are met they are free to create and innovate; they are free to concentrate on being great at their job. When the basic needs of followers are not met, a leader cannot rationally believe that they are getting the best out of their people. As

Maslow (1943) stated, "For the man who is extremely and dangerously hungry, no other interest exists but food" (p. 375). Followers cannot have a sense of belonging, good self-esteem, or in any way self-actualize if they are still stuck searching for the basic safety and physiological needs. This is the reason the servant leader is successful, especially within emergency services organizations (Lichtenwalner 2015). They meet the needs of followers and reap the rewards. Because the servant leader benefits by serving their followers, they receive the gift of trust, and from that trust comes a greater power than that which is demanded (Russell 2016).

References

Bartone, P. T., Roland, R. R., Picano, J. J., & Williams, T. J. (2008). Psychological hardiness predicts success in US Army Special Forces candidates. *International Journal of Selection and Assessment, 16*(1), 78–81.

Castellano, C., & Everly, G. (2006). Rebuilding psychological fences: Reducing trauma through personal and response management. In J. Violanti & D. Paton (Eds.), *Who gets PTSD* (pp. 113–123). Springfield: Charles C Thomas Pub Ltd..

Greenleaf, R. (1970). *The servant as a leader*. Indianapolis: Greenleaf Center.

Greenleaf, R. (1977/2002). *Servant-leadership: A journey into the nature of legitimate power and greatness*. Mahwah: Paulist Press.

Greenleaf, R. (1996). The ethic of strength. In D. Frick & D. Spears (Eds.), *On becoming a servant leader* (pp. 13–99). New York: Jossey-Bass.

Hart, D., & Sussman, R. (2005). *Man the hunted: Primates, predators, and human evolution*. New York: Basic.

Hing, M., Cabrera, J., Barstow, C., & Forsten, R. (2012). Special operations forces and incidence of post-traumatic stress disorder symptoms. *Journal of Special Operations Medicine, 12*(3), 23–35.

Hunter, J. (2004). *The world's most powerful leadership principle: How to become a servant leader*. New York: Crown.

Killgore, W. D., Cotting, D. I., Thomas, J. L., Cox, A. L., McGurk, D., Vo, A. H., & Hoge, C. W. (2008). Post-combat invincibility: Violent combat experiences are associated with increased risk-taking propensity following deployment. *Journal of Psychiatric Research, 42*(13), 1112–1121.

Kimmel, M. (2011). *Manhood in America: A cultural history* (3rd ed.). Oxford: Oxford University Press.

Kirschman, E. (2004). *I love a firefighter: What every family needs to know*. New York: Guilford Press.

Kirschman, E. (2006). *I love a cop, revised edition: What police families need to know*. New York: Guilford Press.

Koran, L. (2016). Special Ops commander tries to lessen the stigma of getting help. Retrieved from https://www.cnn.com/2015/10/06/politics/special-operations-forces-ptsd-mental-health/index.html.

Lichtenwalner, B. (2015). Servant leader company list. Retrieved from: http://modernservantleader.com/featured/servant-leadership-companies-list/.

Maslow, A. (1943). A theory of human motivation. *Psychological Review, 50*, 370–396.

Russell, E. (2014a). *The desire to serve: Servant leadership for fire and emergency services*. Westfield: Robert K. Greenleaf Center for Servant Leadership.

Russell, E. (2014b). Servant leadership theory and the emergency services learner. *Journal of Instructional Research, 3*(1), 64–72.

Russell, E. (2016). Servant leadership's cycle of benefit. *Servant Leadership: Theory & Practice,* *3*(1), 52–68.

Russell, E., Broomé, R., & Prince, R. (2015). Discovering the servant in fire and emergency services leaders. *Servant Leadership: Theory & Practice, 2*(2), 57–75.

Shakespeare-Finch, J. (2006). Individual differences in vulnerability to post trauma deprivation. In J. Violanti & D. Paton (Eds.), *Who gets PTSD* (pp. 33–49). Springfield: Charles C Thomas Pub Ltd..

Shanker, T., & Oppel, R. (2014). War's elite tough guys, hesitant to seek healing. *The New York Times* (pp. A1).

Skipper, L. D., Forsten, R. D., Kim, E. H., Wilk, J. D., & Hoge, C. W. (2014). Relationship of combat experiences and alcohol misuse among US special operations soldiers. *Military Medicine, 179*(3), 301–308.

Volanti, J. (2006). The mind-body nexus: Assessing psychological distress and physiological vulnerability in police officers. In J. Violanti & D. Paton (Eds.), *Who gets PTSD* (pp. 17–32). Springfield: Charles C Thomas Pub Ltd..

Chapter 2
Servant Leadership and the Emergency Services

A noble leader answers not to the trumpet calls of self-promotion, but to the hushed whispers of necessity.

—*Mollie Marti*

Abstract This chapter presents a brief history of the philosophy of servant leadership. It begins by delineating upon its identified characteristics, constructs, and attributes along with its place within the emergency services. Robert K. Greenleaf first penned the modern concept of servant leadership philosophy in 1970. Since that time, the philosophy has spread throughout organizations. Today, there are thousands of books, articles, and research projects, not to mention consulting and coaching centers that teach the philosophy. In addition, universities throughout the world have adopted the philosophy in their management and leadership programs. Recently, empirical works involving servant leadership within the emergency services have emerged and this chapter spotlights them as a way of creating a foundation for understanding its place in the professions.

There exists a power within the philosophy of servant leadership that does not exist in any other leadership approach and it all has to do with a commitment to placing the needs of people within an organization above all else (Greenleaf 1977/2002; Keith 2008). The practice of servant leadership brings out the best in others by meeting their needs, which in turn strengthens the organization and benefits the servant leader (Neuschel 1998; Russell 2016). The approach is one that requires the humbling of one's self (not weakening or lessening) in order to lead others (Hayes and Comer 2010; Nielsen et al. 2010; Patterson 2003). The philosophy transcends all others due to its legitimacy forged in a love for people and practiced through humility (Ferch 2004; Hayes and Comer 2010; Hunter 2004; Patterson 2003). Greenleaf (1977/2002) first theorized the modern philosophy of servant leadership because he desired to give to the world a different approach towards leadership, one that involved realizing success by harnessing the power of people by serving their needs. As Greenleaf (1977/2002) wrote,

© Springer Nature Switzerland AG 2019

E. J. Russell, *In Command of Guardians: Executive Servant Leadership for the Community of Responders*, https://doi.org/10.1007/978-3-030-12493-9_2

Servant leadership begins with the natural feeling that one wants to serve, to serve first. Then conscious choice brings one to aspire to lead. He is sharply different from the person who is leader first, perhaps because of the need to assuage an unusual power drive or to acquire material possessions. For such it will be a later choice to serve after leadership is established. The leader first and the servant first are two extreme types. Between them, there are shadings and blends that are part of the infinite variety of human nature (p. 27).

Greenleaf's (1977/2002) philosophy is made up of three pragmatic questions. The first, "Do those served grow as persons?" (Greenleaf 1977/2002, p. 27), second, "Do they, while being served, become healthier, wiser, freer, more autonomous, more likely themselves to become servants?" (Greenleaf 1977/2002, p. 27), and third, "What is the effect on the least privileged in society, will they benefit or at least not be further deprived" (Greenleaf 1977/2002, p. 27). Greenleaf (1977/2002) argued that positive answers to these questions leads to strengthening a leader's legitimate power through the service of others. Research on the philosophy has found that people become loyal to those that serve them and in-turn, they reciprocate by serving their leaders and organizations (Greenleaf 1977/2002; Hunter 2004; Russell 2016; Spears 2010).

This philosophy revolutionized leadership thought by bringing an approach to the world that is opposite of a desire to lead first (Sipe and Frick 2009). Works pertaining to servant leadership embed one's desire to serve others, to meet their needs. Servant leadership works argue that the path to true happiness comes from the idea of serving others (Greenleaf 1977/2002; Keith 2008). However, unlike current leadership theory, servant leadership bases itself on the notion that the servant leader is one that possesses specific characteristics. These characteristics become focal points for cultivation in those eager to serve others (Spears 2010). Again, the idea of servant leadership is not the weakening of leaders and subduing their legitimacy but rather, strengthening their role through service (Greenleaf 1977/2002), empathy (Spears 2010), and humility (Hayes and Comer 2010). As it is with the emergency services profession, being a servant leader gives meaning to one's life (Keith 2008; Russell 2014).

When writing on the individual, Greenleaf (1977/2002) argued the idea that those who wish to serve transcend to a position of legitimate leadership, which evolves as a byproduct of serving. The philosophy places the individual within society, regardless of stature, in service to the needs of others (Greenleaf 1977/2002). Such a notion involves the improvement and development of others as the cornerstone. Spears (2010) held such a concept, the committing oneself to the betterment and growth of others, as a core characteristic of the servant leader. Here is the individual: impressive, educated and self-aware, committing their life to the service of others for their betterment and in turn, personally benefiting (Greenleaf 1977/2002; Russell, 2016).

The essential component of Greenleaf's (1977/2002) philosophy is the individual and the power they receive from being in the service to others. Greenleaf (1977/2002) posed two specific questions about servant leadership: the first, whether the leader grows as an individual and the second, the growth of the individual served. Greenleaf (1977/2002) places the idea of putting people first as the central tenant of success for an organization, arguing that the institution that puts the needs of its people before all else will in fact realize positive outcomes. Such an idea is based upon the notion that if one's people are taken care of, they in turn will take care of everything else. Greenleaf (1977/2002) wrote that the institution that practices servant leadership flourishes because those served desire to make it so.

The Characteristics of Servant Leadership

It was from the roots of servant leadership that Spears (2010) established what are known as the ten characteristics of the servant leader. Spear's (2010) work took Greenleaf's (1970) writings regarding servant leadership from a theory to a usable and identifiable model based upon ten characteristics identified in the works. Because of Spears (2010) work, the theory of servant leadership now contained specific and measurable characteristics one could use to identify servant leadership qualities within individual leaders.

Derived from an interpretation of Greenleaf's (1970) original essay, the characteristics of the servant leader are listening, empathy, healing, awareness, persuasion, conceptualization, foresight, stewardship, commitment to the growth of people, and building community (Spears 2010). As his work progressed, two additional characteristics were identified—calling and nurturing the joy of spirit, and added to servant leadership characteristics. The characteristics are specific and yet not exhaustive areas to describe the servant leader (Spears 2010). Besides being able to use these characteristics to measure servant leaders, they also function as a way to look inward into one's own leadership characteristics. These characteristics form the foundation of this book.

The Servant Leadership Model

With the expansion of conceptual writings by celebrated authors regarding the philosophy of servant leadership, Farling et al. (1999) argued for the need of empirical studies as well as the development of models and measurement instruments. A major study that followed was conducted by Laub (1999), whose research led to the development of one of the first instruments to be used to study servant leadership known as the *Organizational Leadership Assessment Instrument* (OLA). Developed through a Delphi study, the OLA became a sought-after instrument to assess the presence and amount of servant leadership in organizations (Laub 1999). Contained within Laub's (1999) OLA is the Servant Leadership Model, which identifies six specific areas of measurement for servant leadership: values people, develops people, builds community, displays authenticity, provides leadership, and shares leadership. Since publication, the OLA has led to dozens of published dissertations and theses. Like the characteristics, Laub's (1999) Servant Leadership Model helped shape this work.

The Constructs of Servant Leadership

A groundbreaking work by Patterson (2003) moved beyond the characteristics of the philosophy into the virtues that make up servant leadership. These constructs of servant leadership that derived from Patterson's (2003) work involved identifying specifics within Greenleaf's (1970) servant leadership philosophy. Unlike the 10 characteristics that Spears (2010) established to describe the servant leader, or Laub's (1999) discovery for measuring servant leadership, the seven virtuous constructs embody the theoretical core of Greenleaf's (1970) original essay *The Servant as Leader*. Patterson (2003) argued that the philosophy of servant leadership extended from and beyond the transformational leadership model and therefore demanded its own set of parameters, thus leading to the seven virtuous constructs specific to servant leadership. Patterson's (2003) identified differences led to the creation of seven virtuous constructs specific only to the practice of servant leadership. In order, these constructs are *agapao* (love), humility, altruism, vision, trust, empowerment, and service. Patterson (2003) displayed how each construct flows into the next with the pinnacle resting upon service. It is these constructs that exist within an emergency responder's desire to serve (Russell 2014).

A Place for Servant Leadership in the Emergency Services

The career of the emergency services professional still holds true to the same traditions and passions as those who came before and the love of serving others is still the foundation of what it means to be an emergency services responder (Cortrite 2007; Fleming 2010; Lasky 2006; Morris 1955; Smeby 2005; Wender 2008). Leadership in the emergency services poses a unique set of challenges, where leaders must take two separate, yet simultaneous paths. The first path is leadership associated with on-scene emergency management and control; this situation involves a direct-authoritative role, including giving commands and orders to crews and operators (Anglin 2001; Coleman 2008; Cordner 2016; van Doren 2006). On the emergency scene, a command and control style of leadership is necessary for safe and effective operations (Mitchell and Casey 2007; Smeby 2005). Command and control of an emergency scene is a complex system, where time is of the essence and the environment is one of danger and risk (Bigley and Roberts 2001). Within the emergency services this system needs to exist for safety and order during emergency operations, for the more lethal the task the more ridged the structure of command and control. However, this control has little place outside of emergency response (Kirschman 2004, 2006). This is where the philosophy of servant leadership has its place, not in the command and control of the emergency operations, but rather, within the community of responders.

The second path is leadership apart from an emergency scene, which is a very different role. As stated before, emergency services become a personal identifier for the individuals that operate in the career filed. The majority of fire and EMS personnel work 24–48 h shifts, living with one another in a family-like community. In fact, emergency services personnel commonly refer to their stations as houses. Law enforcement personnel work 12 h shifts relying and interacting with the same partners for years. Therefore, relationships go beyond the stereotypical coworker to a brotherhood/sisterhood (Baker 2011; Cordner 2016; Salka and Neville 2004; Sargent 2006; Seigal 2006; Smith 1972; Smoke 2010).

Servant Leadership and the Emergency Responder

There is a commonalty between the philosophy of servant leadership and the desire to serve that brings the professional emergency services responder to the career field. Specifically, servant leadership and the emergency services responder seemingly share the same virtues: *agapao* (love), humility, altruism, vision, trust, empowerment, and service (Patterson 2003; Russell 2014).

It begins with love. Patterson (2003) placed *agapao* (love) as the first virtuous construct of servant leadership. For the emergency services, it is a love for people that brings the individual to the profession, and it is love that allows one to remain (Cortrite 2007; Lasky 2006). The conscientious decision to enter into an emergency services career comes with an understanding of the inherent dangers associated with the profession (Baker 2011; Salka and Neville 2004). It is a love of serving others in their most vulnerable time of need that has called individuals to the profession throughout history (Morris 1955).

Hayes and Comer (2010) argued that humility is humanity, and for the emergency services responder, that humility shows outwardly with acts of self-sacrifice and care. The emergency services responder deals with others in their most vulnerable situations and in their most critical time of need (Barker 2017; Cordner 2016; Smeby 2005). Through humility, one can reach out to others (Nielsen et al. 2010).

Invited into the life of others, the emergency services responder becomes the humble servant, who when called upon, is willing to give their all (Smith 1972; Useem et al. 2005). Such an act is altruism in its purest form—the giving of oneself for another (Patterson 2003). The altruistic nature of the emergency services profession is one that reaches out to others through a willingness to sacrifice in order to save strangers. Altruism stems from those with a passion to serve others without question and in so doing, willingly gives another one's all (Day 2004).

For the emergency services, vision involves seeing the needs of the community one serves and ensuring those needs are met. Bell and Habel (2009) argued that the visionary rejects complacency and looks towards the future. Inwardly, vision protects the emergency services profession, keeping the career field viable by meeting future needs (Whetstone 2002). The emergency services professional remains committed to being at the ready, which includes taking on different responsibilities for individuals within the organization, as well as the community served (Anglin 2001; Cordner 2016; Fleming 2010).

The nature of emergency service operations is built upon trust. Individuals thrust into emergencies must rely not only on their own abilities, but also the abilities of others (Gilmartin 2002; Klinoff 2012). At the core of the operation is a trust between leaders and followers, as well as coworkers. This trust involves believing in the abilities of those in command to make the right decisions and from this trust, comes a willingness to carry out orders without question (Caldwell et al. 2009). A responder earns trust in the emergency services; it does not come automatically with a position (Baker 2011; Sargent 2006). Instead, it comes over time through one's actions (Caldwell et al. 2009). Furthermore, trust must exist from the leader to the follower, where the actions, commitment, and abilities of the follower allow for the leader to trust them to operate without direct supervision (Caldwell and Hayes 2007).

The trust of others causes a willingness from leaders to empower their followers. Within the emergency services, centralized leadership is a standard practice that seemingly flows over from the emergency scene to the day-to-day activities. Ndoye et al. (2010) argued for removing centralized leadership practices as a way to share decision-making and empower followers. The very nature of the emergency response organization involves multiple independent companies and patrols overseen by junior officers who operate in designated strategic areas (Choo et al. 2011; Cordner 2016; Fleming 2010; Smoke 2010). Companies, patrols and crews are empowered to respond to emergencies and make arrest, tactical, and patient decisions, depending on the size and severity of the situation, free from the direct supervision of chief officers (Coleman 2008; Cordner 2016; Salka and Neville 2004; Smeby 2005). Therefore, the emergency services profession operates in a continuous state of trust and empowerment. Leaders have no choice but to empower their people in order for the service to function.

As Patterson (2003) explained, the constructs come together to form the core construct of service, which Sipe and Frick (2009) argued was the absolute giving of self to the service of others. Service is indeed the core value of the emergency services responder; it is in that essential desire to serve that the individual steps forward. As Greenleaf (1977/2002) wrote, it is from a desire to serve that the leader appears. Therefore, it is from that same desire to serve that the emergency services professional steps forth and through that desire, leads.

Summary

Servant leadership is at the core of the emergency services profession because the constructs that bring forth the individual and formulate their desire to serve are the very same that make up the servant leader (Carter 2007; Cordner 2016; Patterson 2003; Russell 2014). Harnessing that desire to serve and putting it forth as a leadership philosophy for the emergency services holds promise for overcoming poor and destructive leadership habits. Cortrite (2007) found that the practice of servant leadership showed promise for overcoming toxic leadership practices that stifle human relationships and often times lead to destructive work environments within public safety organizations. In addition, as noted earlier, Greenleaf (1977/2002) argued that the practice of servant leadership could overcome toxic practices within organizations, the emergency services being among them.

The philosophy of servant leadership is a completely different way of approaching leadership. For most, becoming the servant leader means being the leader they desired to work for. The philosophy is based upon serving others, not in servitude, but rather, as one who recognizes when the needs of followers are met, they in turn can grow and self-actualize. The philosophy is not "doormat-leadership" because the servant leader is one that sees potential in followers and strives to bring that potential to life. It is a natural fit for the emergency services because it consists of the very constructs of what called you and others to the profession—a desire to serve.

Case Study A

You are a newly appointed assistant chief of the inspection and investigations division for a large metro fire department. You will be overseeing twelve uniformed and sworn personnel and three support staff. One thing you immediately notice upon taking your position is that the members of your division don't seem to get along. In fact, within weeks you have witnessed hostilities, gossip, and passive-aggressive behaviors when interacting with individuals or small groups. Besides this, you also realize that there is a backlog of investigations and commercial inspections that have yet to be completed.

Case Study A Questions

1. How can building a servant leadership culture improve relationships?
2. What specific steps would you take?
3. What do you see as the root of the issue behind the backlog of work?

Case Study B

You were just sworn in as a Major and assigned to command a medium-sized police department narcotics unit. This position comes to you as a lateral transfer from another organization. Before you get your office set up, your new administrative assistant is bringing you up to speed with all the problems and gossip that plague the unit. He mentions that people on the unit dislike your senior detective, that this individual is a bully and sets people up to fail. You have never met this detective or any other members of the unit.

Case Study B Questions

1. What would you say to your administrative assistant?
2. What would be the first thing you would do?
3. What steps would you take to avoid being influenced?

References

Anglin, G. (2001). Company officer training and development-maintaining consistency in a dynamic environment. In *Executive fire officer program*. Emmetsburg: National Fire Academy.

Baker, T. (2011). *Effective police leadership: Moving beyond management*. Flushing: Looseleaf Law Publications.

Barker, K. C. (2017). *Servant leadership and humility in police promotional practices* (Doctoral dissertation, Walden University).

Bell, M., & Habel, S. (2009). Coaching for a vision for leadership: Oh the places we'll go and the things we can think. *International Journal of Reality Therapy, 29*(1), 18–23.

Bigley, G. A., & Roberts, K. H. (2001). The incident command system: High-reliability organizing for complex and volatile task environments. *Academy of Management Journal, 44*(6), 1281–1299.

Caldwell, C., & Hayes, L. (2007). Leadership, trustworthiness, and the mediating lens. *The Journal of Management Development, 26*(3), 261–274.

Caldwell, C., Davis, B., & Devine, J. (2009). Trust, faith, and betrayal: Insights from management for the wise believer. *Journal of Business Ethics, 89*(1), 103–114.

Carter, H. (2007). Approaches to leadership: The application of theory to the development of a fire service-specific leadership style. *International Fire Service Journal of Leadership and Management, 1*(1), 27–37.

Choo, S., Park, O., & Kang, H. (2011). The factors influencing empowerment of 119 emergency medical technicians. *Korean Journal of Occupational Health Nursing, 20*(2), 153–162.

Coleman, J. (2008). *Incident management for the street-smart fire officer*. Tulsa: PennWell.

Cordner, G. (2016). *Police administration*. New York: Routledge.

Cortrite, M. (2007). Servant leadership for law enforcement (Doctoral Dissertation). Available from ProQuest Dissertation and Theses Database (UMI No. 3299572).

Day, C. (2004). The passion of successful leadership. *School Leadership and Management, 24*(4), 425–437.

Farling, M., Stone, A., & Winston, B. (1999). Servant leadership: Setting the stage for empirical research. *The Journal of Leadership & Organizational Studies, 6*(1), 49–72.

Ferch, S. (2004). Servant-leadership, forgiveness, and social justice. In L. C. Spears & M. Lawrence (Eds.), *Practicing servant-leadership: Succeeding through trust, bravery, and forgiveness*. San Francisco: Jossey-Bass.

Fleming, R. (2010). *Effective fire and emergency services administration*. Tulsa: PennWell.

Gilmartin, K. (2002). *Emotional survival for law enforcement: A guide for officers and their families*. Tucson: E-S Press.

Greenleaf, R. (1970). *The servant as a leader*. Indianapolis: Greenleaf Center.

Greenleaf, R. (1977/2002). *Servant-leadership: A journey into the nature of legitimate power and greatness*. Mahwah: Paulist Press.

Hayes, M., & Comer, M. (2010). *Start with humility: Lessons from America's quiet CEOs on how to build trust and inspire others*. Indianapolis: Greenleaf Center for Servant-Leadership.

Hunter, J. (2004). *The world's most powerful leadership principle: How to become a servant leader*. New York: Crown.

Keith, K. (2008). *The case for servant leadership*. Westfield: Greenleaf Center for Servant Leadership.

Kirschman, E. (2004). *I love a firefighter: What every family needs to know*. New York: Guilford Press.

Kirschman, E. (2006). *I love a cop, revised edition: What police families need to know*. New York: Guilford Press.

Klinoff, R. (2012). *Introduction to fire protection*. Clifton Park: Delmar.

Lasky, R. (2006). *Pride and ownership: A firefighter's love for the job*. Tulsa: PennWell.

Laub, J. (1999). Assessing the servant organization: Development of the servant organizational leadership assessment (SOLA) instrument (Doctoral dissertation). Retrieved from ProQuest dissertation and theses database (UMI No. 9921922).

Mitchell, M., & Casey, J. (2007). *Police leadership and management*. Annandale: The Federation Press.

Morris, J. (1955). *Fires and firefighters*. Boston: Little, Brown, & Company.

Ndoye, A., Imig, S., & Parker, M. (2010). Empowerment, leadership, and teachers' intentions to stay in or leave the profession or their schools in North Carolina charter schools. *Journal of School Choice, 4*(2), 174–190.

Neuschel, R. (1998). *The servant leader: Unleashing the power of your people*. East Lansing: Vision Sports Management Group.

Nielsen, R., Marrone, J., & Slay, H. (2010). A new look at humility: Exploring the humility concept and its role in socialized charismatic leadership. *Journal of Leadership & Organizational Studies, 17*(1), 33–43.

Patterson, K. (2003). Servant leadership: A theoretical model (Doctoral Dissertation). Available from ProQuest Dissertation and Theses Database (UMI No. 3082719).

Russell, E. (2014). *The Desire to serve: Servant leadership for fire and emergency services*. Westfield: Robert K. Greenleaf Center for Servant Leadership.

Russell, E. (2016). Servant leadership's cycle of benefit. *Servant Leadership: Theory & Practice, 3*(1), 52–68.

Salka, J., & Neville, B. (2004). *First in, last out: Leadership lessons from the New York fire department*. New York: Penguin.

Sargent, C. (2006). *From buddy to boss: Effective fire service leadership*. Tulsa: PennWell.

Seigal, T. (2006). *Developing a succession plan for United States Air Forces in Europe fire and emergency services chief officers*. Executive Fire Officer Program. Emmetsburg, MD: National Fire Academy.

Sipe, J., & Frick, D. (2009). *Seven pillars of servant leadership: Practicing the wisdom of leading by serving*. Mahwah: Paulist Press.

Smeby, C. (2005). *Fire and emergency services administration: Management and leadership practices*. Sudbury: Jones and Bartlett.

Smith, R. (1972). *Report from engine company 82*. New York: Warner Book.

Smoke, C. (2010). *Company officer*. Clifton Park: Delmar.

Spears, L. (2010). Servant leadership and Robert K. Greenleaf's legacy. In K. Patterson & D. van Dierendonck (Eds.), *Servant leadership: Developments in theory and research* (pp. 11–24). New York: Palgrave Macmillan.

Useem, M., Cook, J., & Sutton, L. (2005). Developing leaders for decision making under stress: Wildland firefighters in the south canyon fire and its aftermath. *Academy of Management Learning and Education, 4*(4), 461–485.

van Doren, J. A. (2006). Leading in a tactical paradise. *Fire Engineering, 159*(1), 12.

Wender, J. M. (2008). *Policing and the poetics of everyday life*. Chicago: University of Illinois.

Whetstone, J. (2002). Personalism and moral leadership: The servant leader with a transforming vision. *Business Ethics: A European Review, 11*(4), 385–392.

Chapter 3
Bureaucracy Within the Emergency Services

Eric J. Russell

With contribution by Rodger E. Broomé

> *The purpose of bureaucracy is to compensate for incompetence and lack of discipline.*
>
> —*James C. Collins*

Abstract This chapter introduces one of the biggest contributors to negativity and toxic work environments within the community of responders: bureaucracy. The bureaucracy has been found to be a dizzying maze-like experience for emergency services responders to navigate. In addition, bureaucracy has been identified as one of the greatest contributors to responder's stress and burnout. This chapter begins with identifying what bureaucracy is, and then moves on to discuss the problems of bureaucracy within the emergency services. At its core, bureaucracy is void of the human experience and relationships. It stands on its own as a structure of policies, procedures, and rules that over time can become more important than the guardians they're supposed to be serving.

Bureaucracy is the structure and process to control assets and people (Weber 1978). The structure itself is rigid and demands conformity. Such rigidity often stifles innovation and imagination, cornerstones of what it means to be human. The bureaucratic is the antithesis of the imaginative. The rigidity of the bureaucratic structure does not allow leaders to apply creativity among people in an organization of any kind within managerial situations (Maslow 1965; Weber 1978).

In the emergency services, the bureaucratic process is there to control the chaos of the emergency scene. The concepts of safety and order on the emergency scene are ingrained into the minds of responders at the start of their professional training and education, and constantly reiterated throughout their career. This process exists to meet emergency objectives that are under time-pressure and consequences. Thus, when it is proposed to have a different leadership approach outside of the emergency scene it is typical to have some objections raised by some in the group who need more structure. These are the individuals who believe that without structural order, anarchy and chaos could result. It is important to proceed with this discussion

© Springer Nature Switzerland AG 2019   21
E. J. Russell, *In Command of Guardians: Executive Servant Leadership for the Community of Responders*, https://doi.org/10.1007/978-3-030-12493-9_3

rationally, but also, with an understanding that it comes from deeply emotional and often irrational places (Maslow 1965). This is a demand for a set of rules and principles in the form of policies and procedures that are written for controlling the future and for anticipating any problem that may arise (May 1991; Mills 1959/2000; Perez et al. 2010; Weber 1978). Maslow (1965) continued pointing out that this is realistically impossible and that the future, by nature, is somewhat unpredictable.

Trying to construct a comprehensive and exhaustive book of rules for any contingency is a futile effort. Maslow (1965) proposes that this is a mistrust of our own self that drives a need to prepare. Maslow (1965) goes on to argue that it is better for most situations that the organization works toward a minimum of the rules rather than to a maximum. Perhaps it is best to regard the size of the book of rules needed as proportional to the size and complexity of the organization of people it is meant to serve. It is worth noting that bureaucracy emerged as a self-created arrangement by human beings. That is, both human will and judgment are governed by predetermined ideas and are made to conform so that freedom of ideas is constrained before it is born. Additionally, governments and organizations have argued for the need of bureaucracy as a way to protect the people (de Vries 2001; Weber 1978).

Bureaucracy and Promotion

Bureaucratic hierarchies frame the employee's world in such a way that makes promotion the only real progress; meaning, the only way for someone to be successful in a bureaucracy is to be promoted (May 1991). Moreover, the pyramid structure of hierarchies creates an ever-diminishing possibility for the employee to promote because each level gets smaller in number (Baker 2011; Seigal 2006). In a competitive-individualist culture, each level of success comes with a winner and multiple losers. Hierarchies deal with this by referring back to the rugged individualist's ethos of, "get back up, dust yourself off, and get to work." If one were to look at the standard distribution (bell curve), one would see that in any group of people performing the same job, there is a right-tail group of excellent performers. On the other hand, there is also a left-tail group who are insufficient performers that need to increase productivity. Finally, in the distribution, there is the middle-group consisting of those performers that hover about the mean. It is from that right-tail group where ideally the next "best man for the job" is coming from. Nevertheless, by promoting only one or two, others who are statistically the same in performance are rejected and pushed back toward the middle-group to try again (Kezar 2001; Vinzant and Crothers 1996).

The mythology of the rugged individual promotes the idea that one gets to the top of the organization through hard work, loyalty, and dependability (May 1991). Yet a scientific-style examination process, which is meant to be unbiased, moderates these objectively, but does not guarantee that the best overall candidate is even securely in the right-tail of the distribution. As a result, decision makers must deal with the dilemma of choosing. The choice is between promoting the person who had the best performance determined by these artificial measures within a one-time context, or the person they believe has been the dependable and efficacious performer over time. More often than some would like to see, there is a functional abandonment of the first choice because it is the qualities of the second choice that adds a contributing member to the next level of the organization. After all, if each rank in the organization has fewer and fewer people, the people need increasingly stronger group members in terms of teamwork and performance (Bruegman 2012; Vinzant and Crothers 1996). These groups must not only be valuable, but they must also insulate themselves in terms of political power because they have ascended in an organization that is competitive at every turn. It's an exclusionary ethos. Once one becomes a winner in a bureaucracy, the fall from the upper levels becomes disastrous because he or she lands among the losers.

At what point does the rugged individual, who keeps being passed over and thus pushed back to the middle to try again, begin to be influenced by that push back? Especially when one considers that each promotional opportunity in bureaucracies is meant to be objective and scientific in its process of validating the winners from among the applicants. The result is that the promotion process begins with application, resume, performance testing, and interviews (Anglin 2001; Iannone et al. 2013). Each audition, if you will, becomes an artificially contrived practice that intentionally ignores the history and past efforts of the candidate. Moreover, it also invites new candidates that did not try out in the last audition, so now the field of competition has changed. Overall, the employee's career is one thing that evolves over time in his or her experience. Nevertheless, at promotion time, the evaluation process becomes a performance snapshot by which the employee's entire history is negated. If it is not entirely dismissed by the bureaucracy implementing resume, performance reviews, and other instruments, his or her historical trajectory as an employee is still moderated. The employee's history is then moderated by the snapshot-style evaluation mode of written exams, role-plays, and interviews—seemingly creating a bait-and-switch.

Bureaucracy in the Emergency Services

The bureaucracy is the container in which the organization exists; it sets the parameters, conditions, and minimal standards. When the capacity of the individual becomes greater than the bureaucratic structure, problems arise, hence the reason to have an organization in a state of constantly developing followers (Keith 2008 p. 43). This bureaucratic empiricism creates problems that affect the emergency services responder as they come to the career with a desire to serve others (Fleming 2010; Russell 2014a, b; Salka and Neville 2004). This is why the bureaucracy is a problem within the emergency services, for as Mills (1959/2000) wrote, "[Bureaucrats] are among the humanistically impoverished, living with reference to values that exclude any arising from a respect for human reason" (p. 106).

Oftentimes the emergency services are compelled to focus on problems rather than focus on solutions. This creates an environment that symbolizes mundane issues causing them to seem the same as real problems and chaos (May 1991). Inconvenience becomes a problem; something missing means things are missing. Followers begin equating labeling the kitchen cabinets or cleaning supply closet to the importance of labeling the paramedic drug-box compartments or evidence kits. The responder begins to believe that they cannot handle exceptions because they believe they might fail to exercise good judgment (Maslow 1965). By mistrusting their abilities as professionals, they imagine that using the bureaucratic approach of policy building ad infinitum will serve them with safety and security from the imagined terrible tragedy that might befall them. However, over burdening people with policies creates a world of servants serving the rulebook instead of the rules serving their essential role. This destructive environment damages responders and unfortunately is common throughout emergency services organizations (Kirschman 2004, 2006).

The bureaucracy leaves people desiring for a far different situation than the one in which they currently find themselves functioning (de Vries 2001). The formal bureaucratic structure works under normal operating conditions; however, the formal must be flexible in order to function within the abnormal situation (Lyden 1974). Vinzant and Crothers (1996) found that those who operate in the field possess a leadership style that holds to their values—values that lead to questioning the need for the bureaucracy. Moreover, responders often need to bypass or even disregard the bureaucratic policy in order to save a life (Henderson and Pandey 2013; Rhodes 2006). Henderson and Pandey (2013) found that paramedics operating on an emergency scene had to ignore policies and protocols in order to save the lives of patients in their care. When writing on emergency services response to hurricane Katrina, Rhodes (2006) says that for the responders to be successful, they needed to rise above the bureaucracy in order to save lives. Furthermore, the bureaucratic structure slows and stifles emergency and disaster planning that makes all the difference to the least fortunate within society (Aryal and Dobson 2011; Henderson 2004). It's worth noting that the least fortunate of society are of great concern to servant leaders (Greenleaf 1977/2002).

Servant Leadership for Overcoming Bureaucracy

Lloyd (2003) argued that a failure to develop leaders within the emergency services is an immoral act. Furthermore, intervening in the aftermath of a traumatic situation with programs such as critical incident stress debriefings may be too late. Instead, there is a need to build a culture of support that strengthens the individual before the incident ever occurs (Gilmartin 2002; McNally et al. 2003; Paton 2005; Paton et al. 2004; Sheehan and Van Hasselt 2003). Therefore, a need also exists to build a leadership culture that supports such practices and that transcends bureaucracy. Included in this culture is meeting the needs of individuals by fostering their intelligence and building a community of social support (McNally et al. 2003; Paton 2005; Paton et al. 2004). That concept is found at the core of servant leadership philosophy (Greenleaf 1977/2002; Patterson 2003; Sipe and Frick 2009; Spears 2010).

The problems that arise between the responder and the bureaucracy go beyond the operational aspect of the profession. Emergency services responders make meaning out of their work; the profession becomes a part of their identity (Jensen 2005). This is also the case for those that assume officer level positions, as their rank adds to their identity. Part of what defines these individuals is their role as emergency services leaders. However, the issue arises when many get promoted to officer positions without ever receiving an education, or even a class on being a leader (Russell 2014a). The research of Taylor et al. (2007) found a need for leadership preparatory programs for followers in order to mold them into the leaders of the future. The rational for finding a pathway for servant leadership within the emergency services has to do with the servant leader being held in much higher regard by followers than others leaders (Taylor et al. 2007). In addition, unlike any other approach towards leadership, the virtues of servant leadership mimic the virtues that define what it means to be an emergency services responder (Carter 2007; Cortrite 2007; Russell 2014a, b). Fostering these virtues into a common leadership approach throughout the emergency services holds promise for undoing bureaucratic practices.

Summary

Understandably, organizations need structure. This structure is commonly referred to as bureaucracy. Policies, procedures, standards, and rules govern the structure and form its existence. The problem is not the bureaucratic structure, but rather, the placing of said bureaucracy above the people. When policies exist so that the people can serve them, the policy is the problem. When the law becomes more important than the spirit of the law, the enforcer of said law is the problem. The argument throughout this book is for servant leadership within the emergency services to serve the people that make up the community of responders. It is not advocating the removing of structures; instead, it is making the case for the structure to serve the people.

Case Study A

On the way back from a brushfire along a remote road in an unincorporated section of a county, a medic-engine company released from a fire happens across an automobile versus bicyclist accident on its way back to the station. A bicyclist has struck the side of an automobile when the vehicle rolled past a stop sign, tossing the rider over the vehicle. The rider had suffered multiple life-threatening injuries and was found unconscious. Due to the remote location, the engine's lieutenant radioed the emergency dispatch center requesting a helicopter. Dispatch informed the lieutenant that both helicopters were assigned to other incidents and they were sending an ambulance with an estimated time of arrival being 30 min one way. Knowing that time was of the essence, the lieutenant then called the emergency room requesting permission to transport the patient via the engine company's personnel compartment. For unbeknownst reasons, the physician on duty did not agree and told the lieutenant to wait for transport to arrive. One of the paramedics then informed the lieutenant that the patient had a collapsed lung, and vitals were crashing. The lieutenant made the decision that waiting would cost the patient's life and disregarded orders from the physician. The patient was packaged on a backboard, loaded in the engine, and they went enroute to intercept the ambulance.

In the aftermath, it was found that the decision to transport the patient against orders saved the patient's life by reducing the time it took to get the patient to a trauma-one center. Nevertheless, the medical director over the department's emergency medical program is calling for disciplinary actions against the lieutenant for disobeying the orders of the on-call physician.

Case Study A Questions

1. In this case, what is the law and what is the spirit of the law?
2. As the chief of department, how would you handle this situation and why?
3. How do the possible outcomes of this incident impact future decisions and operations?

Case Study B

You are the chief of department for a small-rural police department. Standing in front of you is one of your junior patrol officers who is having disciplinary actions taken against her after a late-night traffic stop where she pulled over a late model vehicle with expired tags and a burned-out break light. The vehicle had two occupants, one male driver and one female passenger. When the officer made it over to the vehicle she asked the driver for their license and registration. The individual driving handed the officer an expired registration and an expired license. After running the driver's information through the criminal database, she found that the driver had multiple arrests and several prior convictions for drug possession. She returned to the vehicle and asked the driver to exit the driver's side door and to come with her to the back of the vehicle where they could talk.

The officer noticed that the passenger in the vehicle was an elderly woman on oxygen. The driver explained to the officer that the passenger was his grandmother who has lung cancer and that they were coming from a local emergency room where the passenger was treated for breathing complications. The driver was able to produce paperwork corroborating the story. The driver went on to explain that the vehicle belonged to the passenger and was their only mode of transportation. He was his grandmother's only caregiver and lived with her. In addition, and without being asked, the driver also stated that he had a drug-related criminal history and was going on 1 year being clean.

It is the department's policy that the car be impounded, and the driver issued a citation; however, because of the circumstances, the officer decided to give the driver a warning and let him drive the several miles home. As the vehicle was pulling away, one of your patrol sergeants arrived on scene. The officer explained what happened on the call. The sergeant was furious and demanded the officer be reprimanded for her actions.

Case Study B Questions

1. How would you handle this situation?
2. In this case, what is the policy and what is the spirit of the policy?
3. How does your decision whether to reprimand the officer impact future behaviors?

References

Anglin, G. (2001). Company officer training and development-maintaining consistency in a dynamic environment. In *Executive fire officer program*. Emmetsburg: National Fire Academy.

Aryal, K., & Dobson, O. (2011). A case study from the national disaster management institute in the republic of Korea. *The Australian Journal of Emergency Management, 26*(4), 34.

Baker, T. (2011). *Effective police leadership: Moving beyond management*. Flushing: Looseleaf Law Publications.

Bruegman, R. (2012). *Advanced fire administration*. Upper Saddle River: Pearson.

Carter, H. (2007). Approaches to leadership: The application of theory to the development of a fire service-specific leadership style. *International Fire Service Journal of Leadership and Management, 1*(1), 27–37.

Cortrite, M. (2007). *Servant leadership for law enforcement* (Doctoral Dissertation). Available from ProQuest Dissertation and Theses Database. (UMI No. 3299572)

de Vries, M. (2001). The anarchist within: Clinical reflections on Russian character and leadership style. *Human Relations, 54*(5), 585–627.

Fleming, R. (2010). *Effective fire and emergency services administration*. Tulsa: PennWell.

Gilmartin, K. (2002). *Emotional survival for law enforcement: A guide for officers and their families*. Tucson: E-S Press.

Greenleaf, R. (1977/2002). *Servant-leadership: A journey into the nature of legitimate power and greatness*. Mahwah: Paulist Press.

Henderson, L. J. (2004). Emergency and disaster: Pervasive risk and public bureaucracy in developing nations. *Public Organization Review, 4*(2), 103–119.

Henderson, A., & Pandey, S. (2013). Leadership in street level bureaucracy: An exploratory study of supervisor worker interactions in emergency medical services. *International Review of Public Administration, 18*(1), 7–23.

Iannone, N., Iannone, M., & Bernstein, J. (2013). *Supervision of police personnel*. New York, NY: Pearson.

Jensen, M. (2005). The relationship of the sensation seeking personality motive to burnout: Injury and job satisfaction among firefighters. Retrieved from http://scholarworks.uno.edu/.

Keith, K. (2008). *The case for servant leadership*. Westfield: Greenleaf Center for Servant Leadership.

Kezar, A. (2001). Investigating organizational fit in a participatory leadership environment. *Journal of Higher Education Policy & Management, 23*(1), 85–101.

Kirschman, E. (2004). *I love a firefighter: What every family needs to know*. New York: Guilford Press.

Kirschman, E. (2006). *I love a cop: What every family needs to know*. New York: Guilford Press.

Lloyd, H. B. (2003). *Morale matters*. Memphis: Memphis Fire Department. Retrieved from www.usfa.fema.gov/pdf/efop/efo36355.pfd.

Lyden, F. (1974). How bureaucracy responds to crisis. *Public Administration Review, 34*(6), 597.

Maslow, A. (1965, May). Self-actualization and beyond. Conference on the Training of Counselors of Adults, Chatham.

May, R. (1991). *The cry for myth*. New York: W.W. Norton.

McNally, R. J., Bryant, R. A., & Ehlers, A. (2003). Does early psychological intervention promote recovery from posttraumatic stress? *Psychological Science in the Public Interest, 4*(2), 45–79.

Mills, C. (1959/2000). *The sociological imagination*. New York: Oxford.

Paton, D. (2005). Posttraumatic growth in protective services professional: Individual, cognitive and organizational influences. *Traumatology, 11*, 335–346.

Paton, D., Violanti, J., Dunning, C., & Smith, L. M. (2004). *Managing traumatic stress risk: A proactive approach*. Springfield: Charles C. Thomas.

Patterson, K. (2003). Servant leadership: A theoretical model (Doctoral Dissertation). Available from ProQuest Dissertation and Theses Database (UMI No. 3082719).

Perez, L. M., Jones, J., Englert, D. R., & Sachau, D. (2010). Secondary traumatic stress and burn-out among law enforcement investigators exposed to disturbing media images. *Journal of Police and Criminal Psychology, 25*(2), 113–124.

Rhodes, D. (2006). Katrina: "Brotherhood vs. bureaucracy". *Fire Engineering, 159*(5), 71.

Russell, E. (2014a). *The Desire to serve: Servant leadership for fire and emergency services*. Westfield: Robert K. Greenleaf Center for Servant Leadership.

Russell, E. (2014b). Servant leadership theory and the emergency services learner. *Journal of Instructional Research, 3*(1), 64–72.

Salka, J., & Neville, B. (2004). *First in, last out: Leadership lessons from the New York Fire Department*. New York: Penguin.

Seigal, T. (2006). *Developing a succession plan for United States Air Forces in Europe fire and emergency services chief officers* (Executive Fire Officer Program). Emmetsburg: National Fire Academy.

Sheehan, D. C., & Van Hasselt, V. B. (2003). Identifying law enforcement stress reactions early. *FBI Law Enforcement Bulletin, 72*(9), 12–19.

Sipe, J., & Frick, D. (2009). *Seven pillars of servant leadership: Practicing the wisdom of leading by serving*. Mahwah: Paulist Press.

Spears, L. (2010). Servant leadership and Robert K. Greenleaf's legacy. In K. Patterson & D. van Dierendonck (Eds.), *Servant leadership: Developments in theory and research* (pp. 11–24). New York: Palgrave Macmillan.

Taylor, T., Martin, B. N., Hutchinson, S., & Jinks, M. (2007). Examination of leadership practices of principals identified as servant leaders. *International Journal of Leadership in Education, 10*(4), 401–419.

Vinzant, J., & Crothers, L. (1996). Street-level leadership: Rethinking the role of public servants in contemporary governance. *American Review of Public Administration, 26*(4), 457–476.

Weber, M. (1978). *Economy and society*. Oakland: University of California Press.

Chapter 4
The Call to Serve–The Call to Lead

I think there's no higher calling in terms of a career than public service, which is a chance to make a difference in people's lives and improve the world.

—*Jack Lew*

Abstract This chapter brings to life calling and what it means to be an emergency services responder. It looks at certain areas that become part of the leader's calling as well as the benefit to the leader from serving their followers. When someone is called to a career in the emergency services, this profession, this personal sacrifice, places the individual in harm's way and transcends the notion of a typical job. Being an emergency responder becomes a primary identifier of the guardians who perform the work. That call to serve brought the leaders to the career field; that same calling is the leader's call to lead. In addition, this chapter offers a pathway for emergency services leaders to look inward at their own calling and use it to become effective leaders for the community of responders.

For the most part, individuals from different professional backgrounds fail to see their career as a calling (Wrzesniewski et al. 1997). For the emergency services, this is not the case (Antonellis 2007). Being an emergency services professional often-times becomes the identity of the individual. It is a transcendent experience that defines the individual, impacting their self-identity (Dobrow and Tosti-Kharas 2012). This issue becomes clear when the individual responder plans to retire or decides to change careers (Antonellis 2007). The act of leaving emergency services work is more than just starting a new chapter in one's life, it is actually letting go of who one is (Maslow 1969/1971).

For the emergency services leader it is important to understand this, taking it into account when dealing with responders. The psychological issues associated with being an emergency responder run deeper than those in other professions due to matters of self (Dobrow and Tosti-Kharas 2012). For when the responder has an issue, his or her being has an issue. When the responder cannot solve a problem, his

© Springer Nature Switzerland AG 2019
E. J. Russell, *In Command of Guardians: Executive Servant Leadership for the Community of Responders*, https://doi.org/10.1007/978-3-030-12493-9_4

or her being fails to solve a problem. This identity goes even deeper when it comes to rank. Not only is the individual responder defined by what they do for a living, but also by their position and level of authority (Wrzesniewski 2002). This is the reason why in cases of insubordination or when an individual in a leadership position has their authority called into question, it creates an existential crisis (Maslow 1954; Rivkin et al. 2014). To the individual, their authority is part of who they are. When that is disrespected or ignored, it is personal.

Other professions do not wear T-shirts about their profession. Outside of the emergency and military services, it is not common for people to have statues and swag depicting what they do for a living. Walk through the halls of any fire academy, police academy, or paramedic-program classrooms and think about the esthetics of things hanging on the walls. These symbols reflect who these individuals are as people. Some may see them as meaningless, but the fact is they have to believe in what those symbols mean (Maslow 1943). To bring this home, ask yourself how many professions do you know of that have a specific assigned prayer? The firefighter prayer was written in 1958 by firefighter A.W. "Smokey" Linn. The prayer is read at funerals, tattooed on flesh, and is inscribed on plaques, swag, grave markers, and station walls alike. The firefighter prayer states,

> When I am called to duty, God, whenever flames may rage, give me strength to save some life, whatever be its age. Help me embrace a little child before it is too late or save an older person from the horror of that fate. Enable me to be alert and hear the weakest shout, and quickly and efficiently to put the fire out. I want to fill my calling and to give the best in me, to guard my every neighbor and protect his property. And if, according to my fate, I am to lose my life, please bless with your protecting hand my children and my wife.

This prayer captures the language of the firefighter. It uses words that conjure up images of bravery, sacrifice, and service. These words do matter to responders, they are not superficial or passé. For at the heart of any culture is language. Language is the richest and deepest point of one's being. The notion of culture and the power of language holds true within the emergency services. The profession has its own language. It has its own sense of humor. No matter where you go in the world there is a shared language between emergency responders. To be a part of the career field, to be seen as a member of the profession, you have to speak the language. This leads to a global kinship. The beauty of emergency service leadership is those that are put in charge of the emergency services usually know the language. It is because they themselves, before called into a position of leadership, were called into a position of service. They speak the language; they know what others are going through. This is the reason why emergency responders are assigned to critical incident stress debriefing teams, because responders have a hard time opening up to people that do not speak their language.

The purpose of this chapter is to remind an emergency responder why they do their job. It is to spark reflection in one's mind of the reasons they came to this career. People seem to forget why they do the work after a while. For whatever reason, they lose sight of their purpose, and that has a direct and negative impact on the community of responders. Assignments become mundane. Year-after-year of seeing the human condition causes one to become jaded, eroding one's humanity. The same is true when people are promoted. Within the emergency services, officers tend to get caught up in those things that don't matter. They feel unsatisfied and use terms like "pencil pushers" and "babysitter", forgetting the reason why they pursued the career. Their position isn't mundane, and their role isn't useless. In fact, it is the exact opposite. Their role is vital for they are in command of guardians.

Why Are They Here? Why Are You Here?

Is what brought you here the same thing that brought those you lead to the profession? Is it possible that now you find yourself in an executive level leadership position, you may have forgotten why you desired to do this job in the first place?

Internally reflecting on those questions will bring to life your "why". Asking why will assist you in making the right decisions (Sinek 2011). It will gift you moments of reflection on the very reasons you chose this profession, as well as why those you lead are standing in front of you. Sinek (2011) makes the case for asking "why" before you do anything. For the responder, their "why" is about their inner desire to serve.

The desire to serve begins with a calling to the profession (Hall and Chandler 2005). In their own personal and individual way, the responder is called to a life of service to others. Something happens, be it mentally or spiritually, that calls the individual to the career. It is not by ones choosing, it is not if-not-this-then-that. It is not something one falls into, or a Plan B when something else doesn't seem to work out. For the professional emergency responder, it is a calling to be something bigger than self. It is a calling to serve the stranger in their time of need. A calling where one can say I got this, I will make it better.

The issue of calling is a not a cliché, it is not some tradition or belief older responders try to sell to new recruits. It is the lifeblood of the profession. What the individual responder will be asked to endure, the experiences they will be commanded to have, will bring to life the parts of humanity that so many in society try to ignore, water-down, or simply deny. They will become active participants in tragedy and will need to do it all over again during their next shift. Therefore, if it is not a calling, they will soon seek out work that is more lucrative and much safer to do.

So how does this impact you as a leader? Well, it begins with how you treat your people. It starts by not approaching your people as replaceable workers, but rather, somewhat irreplaceable professionals that make up less than 1% of the population. Yes, that's less than 1% of the population. The nature of the profession and what responders are asked to do makes it quite difficult to simply post help wanted ads in the classifieds. You can't take anyone off the streets and make them a responder. Yes, you can train them to be a responder. You can even dress them in the uniform and bring them on the job. However, they will not last and that has to do with the fact that the profession is bigger than that. And that's because to work and survive within the emergency services, one must be called to the profession.

The Profession as Meaning

The emergency services work only becomes meaningful *when* the responder is called to the work (Duffy et al. 2013). This allows the emergency responder to make meaning out of their work. Not only does the profession define the emergency responder, the work gives them purpose (Dik et al. 2009). When the responder can put someone else's world back together they receive a sense of personal satisfaction and achievement. This is what makes the work meaningful, the notion that they can make a lasting impression and a difference in the lives of others (Duffy et al. 2013). Those who receive the care from the responder will possess a lasting memory of that individual. That means long after names are forgotten, the work of that responder is etched on the psyche of the individual that they served. This has meaning to the emergency responder. This is something that transcends money or authority (Koltko-Rivera 2006).

Like the emergency responder, the emergency services leader has to find meaning in their work. This completes the ongoing cycle of emergency response, forming the ongoing leader-follower relationship (Winston 2003). It is this notion, just as it is with the responder, that the emergency services leader finds meaning and satisfaction in serving a community of responders. The leader finds pleasure in the success of followers. As followers' respond and give aid to the community, the leader realizes happiness from their success. Like the responder finding meaning in service to strangers, ensuring the success of the responders becomes the leader's ultimate goal.

A Calling to Lead

There was a time when it was your calling to serve the public and your team as a responder. Now it is your calling to serve them as their leader. As it was when you were called to serve, you cultivated yourself as a professional responder. You attended classes, received certifications, pursued academic degrees, and took part in honing advanced skills. You practiced and refined these skills throughout your career. Today it is your calling to be a leader; it is up to you to advance your leadership abilities in order to fulfill your calling of being in command. Your calling to lead comes with a whole new set of priorities, responsibilities, and obligations. Your role as a leader means that you have transcended beyond what it means to be an emergency responder (Koltko-Rivera 2006).

Your call to lead comes to you the same as your call to serve. The service that you do is just a different way of being in service to others. When you were an emergency responder your responsibility was to your crew and the public. Now as an emergency services leader your responsibility is to the community of responders so they can continue to be responsible to their crews as well as the public. As previously discussed, your success as a leader is measured by the success of your followers. Your calling to lead removes you from the rubber-meets-the-road operations that take place on a daily basis. Your calling now involves ensuring that those entrusted to you are able to meet the needs of the public and the future obligations of the profession.

The Beneficiary of Your Leadership

Realizing whom your leadership benefits begins with a reflection on the following questions. The first is, what are you and your position for? Are you there for yourself? Are you there for them? Or are you there for both yourself and them? Understand that only you can answer these questions. This reflection on oneself is a personal growing experience. If you find yourself coming up with answers that disturb you, then you know the areas in which you need to work on. It needs to be made clear that a dual benefit for both you and your followers is a healthy thing (Russell 2016). The notion of a dual benefit does two things. First, it avoids the trappings of needing to be altruistic outside of the leader-follower relationship. Second, it removes feelings of guilt from being successful.

So much of the skepticism surrounding servant leadership involves a misunderstanding of the philosophy. It is a belief that servant leadership is one-sided, and that being a servant leader is somehow altruistic. The only place where altruism exists within the philosophy of servant leadership is within the leader-follower relationship (Patterson 2003; Winston 2003). Outside of that relationship, there is a direct and indirect benefit to the individual who decides to be a servant leader (Russell 2016). This is known as servant leadership's cycle of benefit (Russell 2016).

Russell (2016) developed this theoretical concept in order to push back against the skepticism surrounding the philosophy. Part of the problem is that many people see the philosophy of servant leadership as servitude. This cannot be further from the truth. When the leader serves a follower, that follower is then free to pursue their work. When a follower is served, they can be creative and innovative (Conley 2007). That creativity and innovation leads to the follower being successful. Their success always directly and/or indirectly benefits the leader (Russell 2016).

The leader that is contemplating a servant leadership pathway for the future needs to understand that though it is benevolent, it is also self-serving. Coming to terms with this makes it easier to bring the philosophy into "alpha" career fields. When you understand that there is a benefit to you from being a servant leader, you no longer see the characteristics, constructs, and attributes of servant leadership as putting you in a position of servitude. Instead, becoming a servant leader becomes a decision of rational selfishness where everybody involved benefits (Russell 2016).

A Call to Self-Actualization

Serving your followers so that they can self-actualize goes to the heart of servant leadership philosophy. If we reflect back to the foundations of servant leadership, one of its pillars asks, "Do your followers grow and self-actualize as a result of having their needs met?" (Greenleaf 1977/2002 p. 27). The purpose of serving the community of responders is so that the individuals that make up the community can one day self-actualize. In fact, this is the end goal for the emergency services executive. Your calling to lead is all about serving followers so that they themselves can stand on their own as strong, resilient, and capable professionals.

Just about every introductory emergency services officer textbook addresses Maslow's (1943) hierarchy of needs. In fact, you can probably close your eyes right now and envision that multicolored pyramid that shows each of the individual levels. Normally this area is just touched on, and the books never seem to explain why it matters. The following hopefully changes that, starting with a quick refresher.

The hierarchy of needs consists of six tiers: physiological, safety, belongingness, esteem, self-actualization, and self-transcendence (Koltko-Rivera 2006; Maslow 1943). Recently, Koltko-Rivera (2006) identified self-transcendence from the works of Abraham Maslow. It is a different level of both philosophical discussion and individual ability; it involves an individual's ability to transcend beyond self—something only the few can do (Koltko-Rivera 2006; Maslow 1969).

Each tier associated with the hierarchy of needs is not a level unto itself; they are ongoing and constant (Maslow 1943). In fact, even when you are getting a follower to the point of self-actualization, they still have the basic psychological and safety needs that cannot be ignored. It is not like a game where you reach the next level and never have to think about the past. It is easy to see how these tiers fit within the community of responders. The physiological and safety needs, the responders desire to belong to something, the esteem that comes from their work—these are all part of an ongoing process of their growth. As noted before, Maslow (1943) stated, "For the man who is extremely and dangerously hungry, no other interest exists but food" (p. 375). This means that you as a leader have to constantly be striving to meet those basic needs of followers. You as a leader don't have the power to self-actualize for them; it is entirely up to the individual. It is through your service from your position that allows them to move towards self-actualization because you are dedicated to meeting their needs.

It must be made clear that when followers' basic needs are not met, a leader cannot expect to get the best out of their people. For at this level, when the basic physiological and safety needs of one's people are not being met, they cannot have a sense of belonging, self-esteem, or self-actualize (Conley 2007). Moreover, when ones' followers cannot self-actualize they in turn cannot meet the needs of the leader. Always remember that psychological health is correlated to one's basic needs (Maslow 1971). If you as a leader want to be served by those you serve, their psychological health is paramount.

This inability to reach the point of self-actualization creates vulnerabilities for responders. Think of the individual responder as a system that has to be strengthened. The weaker the system, the more vulnerable it is to collapse (Garrett 2012). This means that when you stress the system, in this case the responder, because of that vulnerability, they are at greater risk of collapse. That means burnout, mental health issues, or worse.

When one decides to be a servant leader, they commit to meeting the needs of followers (Greenleaf 1977/2002). For the individual that rationally chooses to be a servant leader, they are keenly aware of the needs of those they serve (Greenleaf 1977/2002; Laub 1999; Spears 2010). When followers' needs are met, they can then self-actualize and are free to create and innovate (Conley 2007). This takes us back to servant leadership's cycle of benefit where the served-followers continually fulfill the leader's needs because they as followers have the opportunity to self-actualize since the servant leader continues to serve their needs.

Winston (2003) identified one of the servant followers constructs as self-efficacy; an individual's ability to perform. Within this relationship, the leader who serves their followers has a commitment to the growth of their people (Spears 2010). That commitment fosters a leader to trust and empower followers and followers to trustingly accept such empowerment (Patterson 2003; Winston 2003). From said trust and empowerment, followers grow as persons (Greenleaf 1977/2002). Out of a follower's growth develops self-efficacy (Winston 2003). It is one's self-efficacy that nurtures one's ability to self-actualize (Conger and Kanungo 1988; McCombs 1986; Maslow 1954; Wilson et al. 2007).

All of this comes together when you as a leader have one of your end goals be the self-actualization of your followers. It begins with an understanding that when responders can reach a level of self-actualization they are mentally healthier (Maslow 1971). The self-actualizing emergency responder is more resilient; their posttraumatic growth correlates to their ability to self-actualize (McNally et al. 2003; Paton 2005). Getting your responders to a point of self-actualization means that you have strengthened the community of responders by serving their needs so that they can grow. That growth leads to self-actualization, thus reducing the need to be prescriptive and reactive with mental health services (Paton 2005). It means when followers self-actualize, it is easier to empower and delegate authority to them so they can make the decisions without direct control from you (Patterson 2003; Russell 2014). This frees you up as a leader to concentrate on the bigger picture, to pay more attention to the community of responders as a whole and the individuals within (Russell et al. 2015). When your followers self-actualize you no longer have a responsibility to look over their shoulders. They've grown as individuals and can handle the situation because you served them in a way that allowed for self-actualization.

Research on Serving the Need of Responders

Russell et al. (2015) set forth to conduct a qualitative research study involving the interpretation of company level emergency services officers regarding the role and characteristics of leadership. Parts of this work are presented with permission from the journal *Servant Leadership Theory & Practice* in this and several other chapters as a means of strengthening the servant leadership discussion. One of the attributes that emerged to support their theoretical finding of the notion that emergency services leaders must serve the needs of followers was that leaders must meet the needs of followers (Russell et al. 2015).

The setting of the Russell et al. (2015) study took place at a large metropolitan fire and emergency services organization in the western United States. To conduct the study, the researchers employed a grounded theory design. Grounded theory is a systematic approach of data collection and analysis, which leads to theoretical discovery (Glaser and Strauss 1999). Data collection consisted of 15 questionnaires obtained from uniformed and sworn fire and emergency services officers. Their research findings resulted in a greater understanding of how fire and emergency services company officers perceive and interpret leadership, as well as the development of propositions for further study (Glaser and Strauss 1999). Presented here are the results of their study, specifically, the attribute leaders need to meet the needs of followers. Below are the results of the "leaders must meet the needs of followers" section of the study in the words of the professional responders that took part in the research (Russell et al. 2015).

Meet the needs of followers P1 claimed that leaders needed to "provide followers with what they need to overcome challenges" (P1). P2 argued that, "a leader should have the crew's well-being in mind in all things" (P2). P3 stated that a leader's responsibility involves "facilitating the support to the firefighters so that they may safely, efficiently and effectively deliver emergency service" (P3). P3 expounded by stating that leadership involves the "support of others in the accomplishment of a common task" (P3). P3 identified that "servant leadership is built into responders; a leader needs to realize that a lot of it just needs to be brought out of people" (P3). P4 stated that a leader "acquires the things necessary for the crew to do their job" (P4). P4 then went on to state that "leadership is what more or less sets the standard for the future of the fire service, a leader can be a follower and a follower can be a leader" (P4).

P5 discussed that "leaders need to make the crews feel like they are there for them, and that they have their backs—going to bat for employees that are having problems" (P5). P5 went on to say that this was about "fostering an environment where the employees feel important and protected" (P5). This is accomplished by "encouraging employees to do the right thing to make every situation better" (P5). P6 suggested that leaders have a responsibility to "ensure that people are operating as safely as possible and that people have the resources needed to accomplish their tasks" (P6). P7 said that leaders needed to "help followers communicate better, respectfully both up and down the chain" (P7). P7 went on to say that leaders must "know the needs of our citizens, what are their needs and how can we serve them, relating that to members and listen to how they would address issues" (P7).

P8 argued that a leader must "make sure self and crews are ready to act and make sure it happens appropriately" (P8). P9 stated that a leader is responsible for keeping the "crew happy and mentally stable" (P9). P11 discussed specific traits a leader needed to have in order to successfully meet the needs of followers—"to be calm, wise and knowledgeable, a leader should also be approachable, non-judgmental" (P11). Adding to this idea, P12 said that a leader must be "knowledgeable and have the ability to influence persons below him or her to believe and follow" (P12). To meet the needs of followers, P12 stated that, "a leader works with the group and for the group" (P12). Meeting the needs of followers comes from a leader desiring "to do good and care about individuals" (P13). According to P14, when you meet followers' needs, leaders "empower their people" (P14). P15 stated that leaders "needed to get what followers need" (P15). P15 went on to state that a leader "should be worrying about followers needs before your needs, because in the end they're the ones doing the work" (P15).

Meeting the needs of followers goes to the heart of what it means to be a servant leader (Greenleaf 1977/2002). As Greenleaf (1977/2002) argued, "The difference manifests itself in the care taken by the servant-first to make sure that other people's highest priority needs are being served" (p. 27). When the follower's needs are met, they in turn can self-actualize and serve the needs of the organization (Maslow 1943).

Summary

Now do you recall why you became an emergency responder in the first place and how it led you to your leadership position? The profession is a calling. It is a defining part of ones being. Today you find yourself in an executive level leadership position, yet that's not how it started. You were called to serve; you desired to become the guardian, the one people turned to in their time of need. As you continue in your position as an executive level leader, never forget that your followers are in their positions for the very same reasons that you took your position so long ago. Though you no longer serve as a rank-and-file emergency services responder, you are still in absolute service to those who do.

As an emergency services leader you are responsible for the future of the profession. You are called upon now to serve and cultivate the future leaders. It is they who one day will sit in the very seat that you do today. Therefore, they have to grow and they have to self-actualize. Self-actualization is imperative for the responder to be able to let go of self and serve others (Maslow 1943). If their needs are not met they can't be expected to carry the burdens of the work; moreover, they cannot be expected to ever assume healthy leadership roles. The weight of not having their physiological or safety needs met can be crushing—the emergency services leader is held accountable for this.

You as the servant leader, simply by being that which you are called to be, fosters an individual's ability to grow and self-actualize (van Dierendonck and Nuijten 2010). According to Jones and Crandall (1986), "For an individual to actualize or make real that which is discovered, there must be a sense of control of one's destiny and a commitment to one's own principles and values" (p. 70). The philosophy of servant leadership shares the same principles and values of the responder, and it begins with a call to serve (Carter 2007; Russell 2014; Russell et al. 2015).

Case Study A

You have just been appointed the deputy chief of operations over emergency medical services for a large metro fire department. For the last 2 years, your organization has been under investigation for narcotics abuse and theft of opioid medications.

Seven of your firefighter paramedics, one lieutenant, and one battalion chief have been terminated. This incident has been all over the local television news stations, written about in national trade publications, and is an ongoing topic of articles in the largest newspaper in your area.

Case Study A Questions

1. What is the impact of this incident on your followers within the organization?
2. As the newly appointed executive level leader over this division, how do you help bolster your followers' self-esteem?
3. What future implications do you think this will have on your organization?

Case Study B

As the newly appointed chief of police for a medium sized city with a force of 72 officers, your new department faced several rounds of cuts to both pay and benefits. On average, each officer has seen their paychecks reduced by 15% and medical out-of-pocket expenses increased by 10%. Those on the lower rungs of the rank structure have taken on part-time jobs to make ends meet. In addition, on your first day you are made aware that the number of officers laterally transferring to other departments in the last twelve months is 11.

Case Study B Questions

1. What is the impact of loss of pay and benefits on your organization?
2. What is the first thing you would do as chief?
3. What steps would you put into place to avoid this in the future?

References

Antonellis, J. J. (2007). Coping with the challenges of forced retirement. *Fire Engineering, 160*(8), 91–100.

Carter, H. (2007). Approaches to leadership: The application of theory to the development of a fire service-specific leadership style. *International Fire Service Journal of Leadership and Management, 1*(1), 27–37.

Conger, J. A., & Kanungo, R. N. (1988). The empowerment process: Integrating theory and practice. *Academy of Management Review, 13*(3), 471–482.

Conley, C. (2007). *Peak: How great companies get their mojo from Maslow.* Hoboken: Jossey-Bass.

Dik, B. J., Duffy, R. D., & Eldridge, B. M. (2009). Calling and vocation in career counseling: Recommendations for promoting meaningful work. *Professional Psychology: Research and Practice, 40*(6), 625–632.

Dobrow, S. R., & Tosti-Kharas, J. (2012). Listen to your heart? Calling and receptivity to career advice. *Journal of Career Assessment, 20*(3), 264–280.

Duffy, R. D., Allan, B. A., Autin, K. L., & Bott, E. M. (2013). Calling and life satisfaction: It is not about having it, it is about living it. *Journal of Counseling Psychology, 60*(1), 42–52.

Garrett, T. J. (2012). Can we predict long-run economic growth? *Retirement Management Journal, 2*(2), 53–61.

Glaser, B., & Strauss, A. (1999). *Discovery of grounded theory: Strategies for qualitative research.* Piscataway: Adline Transaction.

Greenleaf, R. (1977/2002). *Servant-leadership: A journey into the nature of legitimate power and greatness*. Mahwah: Paulist Press.

Hall, D. T., & Chandler, D. E. (2005). Psychological success: When the career is a calling. *Journal of Organizational Behavior, 26*(2), 155–176.

Jones, A., & Crandall, R. (1986). Validation of a short index of self-actualization. *Personality and Social Psychology Bulletin, 12*(1), 63–72.

Koltko-Rivera, M. E. (2006). Rediscovering the later version of Maslow's hierarchy of needs: Self-transcendence and opportunities for theory, research, and unification. *Review of General Psychology, 10*(4), 302–317.

Laub, J. (1999). Assessing the servant organization: Development of the servant organizational leadership assessment (SOLA) instrument (Doctoral dissertation). Retrieved from ProQuest Dissertation and Theses Database (UMI No. 9921922).

Maslow, A. H. (1943). A theory of human motivation. *Psychological Review, 50*, 370–396.

Maslow, A. H. (1954). *Motivation and personality*. New York: Harper.

Maslow, A. H. (1969). The farther reaches of human nature. *Journal of Transpersonal Psychology, 1*(1), 1–9.

Maslow, A. H. (1971). *The farther reaches of human nature*. New York: Viking.

McCombs, B. L. (1986). The role of the self-system in self-regulated learning. *Contemporary Educational Psychology, 11*(4), 314–332.

McNally, R. J., Bryant, R. A., & Ehlers, A. (2003). Does early psychological intervention promote recovery from posttraumatic stress? *Psychological Science in the Public Interest, 4*(2), 45–79.

Paton, D. (2005). Posttraumatic growth in protective services professional: Individual, cognitive and organizational influences. *Traumatology, 11*, 335–346.

Patterson, K. (2003). Servant leadership: A theoretical model (Doctoral Dissertation). Available from ProQuest Dissertation and Theses Database (UMI No. 3082719).

Rivkin, W., Diestel, S., & Schmidt, K. (2014). The positive relationship between servantleadership and employees' psychological health: A multi-method approach. *Zeitschrift Für Personalforschung, 28*(1), 52–72.

Russell, E. (2014). *The desire to serve: Servant leadership for fire and emergency services*. Westfield: Robert K. Greenleaf Center for Servant Leadership.

Russell, E. (2016). Servant leadership's cycle of benefit. *Servant Leadership: Theory & Practice, 3*(1), 52–68.

Russell, E., Broomé, R., & Prince, R. (2015). Discovering the servant in fire and emergency services leaders. *Servant Leadership: Theory & Practice, 2*(2), 57–75.

Sinek, S. (2011). *Start with why*. New York: Penguin Group.

Spears, L. (2010). Servant leadership and Robert K. Greenleaf's legacy. In K. Patterson & D. van Dierendonck (Eds.), *Servant leadership: Developments in theory and research* (pp. 11–24). New York: Palgrave Macmillan.

van Dierendonck, D., & Nuijten, I. (2010). The servant leadership survey: Development and validation of a multidimensional measure. *Journal of Business and Psychology, 25*(10), 1–19.

Wilson, F., Kickul, J., & Marlino, D. (2007). Gender, entrepreneurial self-efficacy, and entrepreneurial career intentions: Implications for entrepreneurship education. *Entrepreneurship Theory and Practice, 31*(3), 387–406.

Winston, B. (2003). Extending Patterson's servant leadership model: Explaining how leaders and followers interact in a circular model. Regent University Servant Leadership Roundtable Regent University.

Wrzesniewski, A. (2002). "It is not just a job": Shifting meanings of work in the wake of 9/11. *Journal of Management Inquiry, 11*(3), 230–234.

Wrzesniewski, A., McCauley, C. R., Rozin, P., & Schwartz, B. (1997). Jobs, careers, and callings: People's relations to their work. *Journal of Research in Personality, 31*, 21–33.

Chapter 5
The Responder's Servant Listener

> *Don't assume, because you are intelligent, able, and well-motivated, that you are open to communication; that you know how to listen.*
>
> —*Robert K. Greenleaf*

Abstract This chapter discusses the successful emergency services leader as one that strives to become a servant listener. A servant leader is one who is a great listener. Servant leadership begins with listening, becoming aware of what is being said and what's not being said. Listening is so much more than simply hearing. It involves being open to knowledge, it is about not being afraid of the truth. The servant listener is an empathetic listener that desires an organizational structure built upon open dialogue. Listening is no longer a soft skill, but rather, one that takes practice because it defines a leader's character.

Have you ever had a conversation with somebody and realized that when you are talking you can tell they are not listening to you? You know just by their face, eyes, and their body language that all they're doing is waiting for their turn to say something. They hear nothing you say; they're not listening to you. This is a frustrating situation, and unfortunately, within organizations and relationships, it is a common occurrence. Some of this has to do with attitude, some ego, and some of it has to do with the fact that people simply lack an ability to listen to others.

When you review after-action reports within the emergency services, it is commonplace to discover that just about every report will have the exact same finding. They all seem to point to a failure of communication. If you ever tried sending information around a "campfire circle" exercise, you'll understand the breakdown in communication. In that situation, you'll have a dozen or so people sitting around the circle. The first person leans over to the individual on their left and tells them a statement. It is then up to that individual that heard the statement to pass it along so it moves around the circle. More times than not when that information gets back to the original person, the statement has completely changed. This example goes to the heart of our inability to listen to what is being said to us. Unfortunately, that's just the tip of the iceberg. The real problem is not simply failing to listen to the things being said, but also, completely missing what is not being said.

The ability to listen is an art (Back 2007). Listening is a learned skill. Leaders that develop the ability to listen will thrive in their position (Thill et al. 1991). When a leader possesses the ability to listen, they hear the needs of their people (O'Connor and Flin 2003). All too often because it is one of the senses of humans, listening is thought of as a given—equivalent to hearing (Borg 2010). However, there is a fundamental difference between hearing and listening. Hearing is simply biology; listening is a honed skill (Burgoon et al. 1994).

When Spears (2010) identified the characteristics of servant leadership, he placed the ability to listen first. The servant leader begins with listening; because they understand that unless they know the truth there is no way that they can lead. Without listening, the leader sets forth on a journey in the wrong direction. Clueless to what's really going on, the poor-listening leader is essentially ineffective. This pitfall is seen within the emergency services when an individual assumes command of an organization that they were never a part of, appointed from the outside. They decide to come into the organization and hit the ground running. Those individuals usually use statements such as "there's a new sheriff in town" or "it's different now that I'm here". Such behavior leads to failure right from the start: it is a learned habit that is dripping with ego.

The servant leader understands that in order to be successful they need to spend a lot more time listening and a lot less time talking. There is no way that you could possibly fix a problem unless you know about the problem and its underlying causes. As a skilled listener you also know that what's not being said is just as important as what's being said. Moreover, you also know that body language, mannerisms, and behavior all add to the story.

Think about having a conversation with somebody who is frustrated. You can tell by their body language, the way that they move or pace (Borg 2010). You can see facial expressions, the way that they stand; these things tell you something is wrong. The servant leader understands this, they don't egg on one's frustration, nor do they push someone to the breaking point. Instead, the servant leader empathetically deescalates the situation. For at that moment, when your follower is in front of you and frustrated, they are not in a proper emotional state to deal with the situation. Humbly, the leader has to step back and serve that individual in their time of obvious need.

For successful emergency responders, the skill is already there. They are required to perform it on the emergency scene. When the situation tempo is heated, when the individual victim or their family is irate, aware responders understand that not taking a moment to calm the situation can lead to tragedy or somebody going away in handcuffs. Successful emergency responders have honed this skill. Now, it needs to become commonplace in leadership situations outside of the emergency response, because no matter how professional a responder is, they are human, and humans get frustrated. And so much of one's frustration is based upon miscommunications, not listening, and not being listened to.

Listening as a Hard Skill

The art of leadership and the ability to listen are all too often considered soft skills (Clarke 2016). Simply by using the word soft to describe the skill psychologically sets the tone that it is a "weak" skill. And by labeling it a weak skill, it becomes easily swept aside in the face of harder skills such as bravery and strength. Using the word soft does a disservice to the vital importance of being able to listen. For some, it creates a stop-barrier (Hulbert 1989).

Believing that listening is a soft skill is a misconception of the characteristic. In fact, since communication failures are always front-and-center with tragedies, it is disingenuous to believe that listening is soft skill. Not only is listening a hard skill that needs to be practiced and cultivated, it is a highly sought-after characteristic in top organizations (Clarke 2016; Jackson and Jensen 2008). Along with an ability to think critically and make decisions, organizational leadership desires their people to be capable and effective listeners (Clarke 2016; Jackson and Jensen 2008).

Since the art of listening is sold as a soft skill, officer programs within the emergency services seemingly brush over the topic of listening within its communications portion. If one just uses after-action reports and published works regarding responder burnout, it becomes easy to justify increasing training and education regarding listening for individuals studying company officer level material (Lloyd et al. 2015; Russell 2014). It is up to leaders within the emergency services profession to tear down this barrier. Ignoring the need to cultivate an individual's ability to listen because other things take precedence is no longer a viable argument. That's the best part about the data that exists: it points directly to the need for effective listening and communications within emergency services (Pandey et al. 2016).

Research has identified that followers perceive listening as one of the desired characteristics of an emergency services leader (Russell et al. 2015). This finding comes directly from those in the profession, the very individuals who exist where the rubber-meets-the-road—they desire their leaders to be strong listeners (Russell et al. 2015). Therefore, if the data corroborates it, the after-action findings justify it, and those in the field desire it, then listening needs to be at the forefront of leadership development within the emergency services. And going forward, listening can no longer legitimately be classified as a soft skill for emergency services leadership.

The nature of emergency services work can have unintended consequences on an individual's personality. Over time, dealing in human tragedies, being lied to by individuals on a scene, seeing what people are capable of doing to themselves and others, can lead to a hardening of the heart. Since the emergency services promote from within, those past experiences as an operator come with the individual into their role as a leader. As it has been addressed before in this work, experiences cannot be undone and what one sees cannot be unseen. For the emergency services leader, it is very important for them to understand that even though they cannot erase their experiences, they can reflect on them with the hopes of growing from them. This matters because that baggage creates a barrier to being able to listen with an open heart (Hulbert 1989).

Gearhart and Bodie (2011) argued that one's ability to listen with an open heart is a social skill needed for healing. The leader who can find a way to listen with an open heart overcomes feelings of not being heard (Myers 2000). Listening with an open heart allows the leader to give of themselves to their followers (Myers 2000). When followers realize that their leader is listening to them, it changes the dynamics of the leader-follower relationship. No longer do followers feel that they're not being heard, or what they're saying doesn't matter. Listening with an open heart means that a leader is actively trying to hear what is being said, as well as what's not being said. The leader who listens with an open heart can sift through the jargon and identify the pain. Oftentimes, what is not being said is what the leader really needs to hear; however, if they can't listen with an open heart, then they fail to recognize the vital truths that often come from silence (Myers 2000).

Research on Listening Within the Emergency Services

Addressed previously in chapter four, the research of Russell et al. (2015) involved the interpretation of company level emergency services officers regarding the role and characteristics of leadership. One of the attributes that emerged to support their theoretical finding of the notion that emergency services leaders must serve the needs of followers was that leaders must listen to followers (Russell et al. 2015). Presented here are the results of their study, specific to the attribute that leaders must listen to followers. Below are the results of the study in the words of the professional responders that took part in the research.

Listens to followers P1 argued that through listening, leaders attempt to "understand the challenges followers face" (P1). P15 stated that, "leaders need to be better listeners than talkers" (P15). P2 said that "this requires a culture of open communication; it also requires that systems be evaluated to allow for that communication" (P2). P3 discussed that listing was done to "keep your finger on the pulse of the boots on the ground" (P3). P4 stated that a good leader "must ask questions and truly attempt to understand the followers concerns and voices" (P4). P5 added that leaders must "visit the stations and talk with the crews" (P5). P7 also stated that, "leaders need to visits the stations as well as the supporting departments" (P7). This was also the case for P8, who discussed that "heeding feedback from followers and conducting station visits" was necessary (P8). P5 elaborated more on this idea by arguing that it was about "asking for feedback and being open to suggestions; crews need to feel like the leaders at the top think about how their decisions affect those people fighting fire and touching patients" (P5). P6 stated that, "leaders need to have an open-door policy" (P6). P9 argued that leaders "must make a point to listen to followers" (P9). P10 stated that, "executive level leaders need to stay engaged in all levels of the department through daily interaction" (P10). P11 discussed that it is important to have "face-to-face conversations, leaders have to be willing to listen, not just hear" (P11). P12 lobbied for "open forums and daily interaction, they seem to be a better way to understand what the organizations employees feel" (P12). P13 stated interaction allows a leader to "be a part what goes on everyday" (P13). According to P14, this is accomplished by "meeting with all crews multiple times throughout the year" (P14).

The Servant Listener

To be a servant leader one needs to be a servant listener—the two are one in the same (Spears 2010). So how does one become a servant listener? It begins with your desire to serve (Frick 2011; Greenleaf 1977/2002). It flows from that desire to serve the needs of others. In order to fully understand these needs, one has to listen to the community of responders.

The servant listener listens with an open heart. They are receptive to honest feedback from followers as long as it is done in a respectful way. This is an important aspect to address because the emergency services are built upon a paramilitary structure. Followers must respect the rank of an individual when addressing them. In turn, the higher-ranking authority, when being addressed respectfully, must be open to hearing the truth. Again, disrespectful communication on one side or pride associated with rank on the other side creates barriers to listening (Hulbert 1989; Lloyd et al. 2015).

The servant listener takes the time to hear the things that are being said and remains consciously aware of the things that aren't being said (Laub 1999; Spears 2010). The servant listener avoids the pitfalls of thinking about what they want to say next while individuals are talking to them. In addition, servant listeners set aside bias and openly listen to the facts. They don't downplay the opinions or feelings of the follower; instead, they recognize the human being in front of them and empathetically listen (Gearhart and Bodie 2011).

Only the individual can develop their ability to listen. It is up to them to take the time to cultivate the skills necessary to become a servant listener. It starts with reflecting on what kind of leader one desires to be. Do you desire to be a leader in the know, one whose followers come to them with information and problems? Your role as a servant listener privies you to knowledge other leaders don't have and haven't earned. It all comes from your followers trusting that you are going to listen to them and take what they have to say seriously. It means that you as the leader will not be dismissive of what they have to say. In the end, this action only strengthens your role as a leader and reduces the guesswork (Frick 2011).

After your reflection, it is then up to you to go out among your people and listen to the things being said and learn to be aware of the things that aren't being said. This involves less talking and more listening. It begins with coming out of your office and going to where your people are. Engage them in conversation and learn about them. All too often the daily mundane, the bureaucracy that can wait, takes priority over one's people. If you are a leader confined to an office, then you are not a leader at all. Leadership is about people, and your people are what ultimately matter. If you stay in your office there is no way that you can listen to your people and over time, they will begin to feel as if they don't matter.

Staying in an inner circle of high-level officers and spending your time in an executive suite creates an ivory tower, one where the only people you will hear from are a select few. When you reach a position of executive level leadership, the majority of emergency responders would never dare come to your office even though you say you have an "open door policy". Think about how you felt when you were a young responder. Going to the chief's office was like going to the snake pit. To be effective, you have to be out and about. Becoming a servant listener is about letting go of the "open door policy" and instead having an "open ear, open heart policy" that hears the needs of your people as you spend time among them.

Remember, just because you decide to listen more, does not mean that you are reducing your position or your rank. In fact, it is quite the opposite. By listening, you are gaining power through trust. By listening to your followers, you strengthen the leader-follower relationship.

Summary

When followers believe that their leader hears them, the leader is gifted legitimate power (Greenleaf 1977/2002; Walker 1997). It comes from one's followers being open and honest in their communications with their leader. The leader grows stronger because they have knowledge of what's going on. They have the ability to strengthen their organization because they are listening to their people. And because of that listening, they can heal wounds and fix problems that they had no idea existed. Leaders and followers gain mutual trust from having a relationship built upon listening.

Think about the frustration that comes from not being heard. Think about the impact that it has had on you in the past. How many times have you found yourself in a situation where your leaders were saying things that you had said in the past? They were dismissive at the time, and yet now, they speak as if it is some sought of genius epiphany. Can you remember standing there in front of them, as an expert in the subject they were dismissing? This happens all too often within the emergency services. Your followers are being paid to be professionals; they are well-trained, educated, and highly skilled individuals. They are the experts and want to be treated as such. Followers want to scream, as did you back in the day, when their leaders fail to listen. The problem of one's inability to listen all comes down to this: the leader who doesn't understand how to listen disrespects and dismisses his or her followers as professionals (Lloyd et al. 2015; Walker 1997).

Case Study A

You are a newly appointed chief of emergency medical services for a large federal government operation. For the last 3 years, your organization has faced staffing cuts due to federal spending deficits. The department that you are in charge of operates advanced emergency medical services on three geographically separated installations. The main facility is your largest operation, staffing six paramedic ambulances across the complex. Most of your time is spent in your office dealing with paperwork and policies, and you find it impossible to get out from behind your desk to be with your people. Since taking command, you are hearing that your operations are having problems. Six months after being sworn into your new position, you begin to see grievances come across your desk involving overtime issues, promotions, and training opportunities.

Case Study A Questions

1. What do you think the underlying problem is?
2. What pitfalls have you fallen into as an executive level leader?
3. What changes will you make as a leader going forward?

Case Study B

You are the deputy chief of police over operations for a state police agency. For the last 10 months, the number of arrests and citations resulting from traffic stops has dropped over 40% along stretches of interstate running through a specific county. One year ago, the county elected a new district attorney that had an anti-police stance. Since his swearing in, convictions have plummeted and seemingly guilty parties have walked free. Your troopers seem frustrated when you talk to them; however, all you are getting is the same information every time—that crime just seems to be down.

Case Study B Questions

1. What do you think the underlying problem is?
2. What is it you're not hearing from your troopers?
3. What steps would you take to fix this problem?

References

Back, L. (2007). *The art of listening*. Oxford: Berg.

Borg, J. (2010). *Body language: 7 easy lessons to master the silent language*. Upper Saddle River: Pearson Education.

Burgoon, M., Hunsaker, F. G., & Dawson, E. J. (1994). *Human communication*. Thousand Oaks: Sage Publications.

Clarke, M. (2016). Addressing the soft skills crisis. *Strategic HR Review, 15*(3), 137–139.

Frick, D. (2011). *Greenleaf and servant-leader listening*. Westfield: Greenleaf Center For Servant Leadership.

Gearhart, C. C., & Bodie, G. D. (2011). Active-empathic listening as a general social skill: Evidence from bivariate and canonical correlations. *Communication Reports, 24*(2), 86–98.

Greenleaf, R. (1977/2002). *Servant-leadership: A journey into the nature of legitimate power and greatness.* Mahwah: Paulist Press.

Hulbert, J. E. (1989). Barriers to effective listening. *Business Communication Quarterly, 52*(2), 3–5.

Jackson, M., & Jensen, J. (2008). *Listening is an essential skill for the 21st century.* New York: New York Press.

Laub, J. (1999). Assessing the servant organization: Development of the servant organizational leadership assessment (SOLA) instrument. (Doctoral dissertation). Retrieved from ProQuest Dissertation and Theses Database (UMI No. 9921922).

Lloyd, K. J., Boer, D., Keller, J. W., & Voelpel, S. (2015). Is my boss really listening to me? The impact of perceived supervisor listening on emotional exhaustion, turnover intention, and organizational citizenship behavior. *Journal of Business Ethics, 130*(3), 509–524.

Myers, S. (2000). Empathic listening: Reports on the experience of being heard. *Journal of Humanistic Psychology, 40*(2), 148–173.

O'Connor, P., & Flin, R. (2003). Crew resource management training for offshore oil production teams. *Safety Science, 41*(7), 591–609.

Pandey, N., Pandey, A., & Kothari, D. A. K. (2016). Soft skills in defense services: Need of the hour. *Macro and Micro dynamics for Empowering Trade, industry and Society,* 161.

Russell, E. (2014). *The desire to serve: Servant leadership for fire and emergency services.* Westfield: Robert K. Greenleaf Center for Servant Leadership.

Russell, E., Broomé, R., & Prince, R. (2015). Discovering the servant in fire and emergency services leaders. *Servant Leadership: Theory & Practice, 2*(2), 57–75.

Spears, L. (2010). Servant leadership and Robert K. Greenleaf's legacy. In K. Patterson & D. van Dierendonck (Eds.), *Servant leadership: Developments in theory and research* (pp. 11–24). New York: Palgrave Macmillan.

Thill, J. V., Bovée, C. L., & Cross, A. (1991). *Excellence in business communication.* New York: McGraw-Hill.

Walker, K. L. (1997). Do you ever listen: Discovering the theoretical underpinnings of empathic listening. *International Journal of Listening, 11*(1), 127–137.

Chapter 6
The Responder's Empathetic Healer

I think we all have empathy. We may not have enough courage to display it.

—*Maya Angelou*

Abstract This chapter discusses the role of the emergency services leader as being an empathetic healer to the community of responders. The empathetic healer is one who willingly binds the wounds of his or her followers. The leader's empathy allows them to see their people as people, to not miss the human. The empathetic healer humbly approaches followers, is keenly aware of their needs, desires to know their pain, and sees their role as one that brings the guardians back to the fight. Empathy is a leader's humility; it is their ability to set aside rank and position in moments when it is more important to stand beside one's followers rather than above.

To look upon another with empathy is not the same as looking down on someone. Instead, it is about seeing them through a lens of compassion (Katz 1963). Empathy allows leaders to see past the superficial, the unique, and instead see followers as people. Furthermore, empathy allows one to see the person underneath the pain. Though society does a great job at pretending there is a "normal" and "just fine", the empathetic leader, being keenly aware that there isn't a "normal" nor a "just fine" sees, accepts, and loves people for who they are. In addition, empathy fosters a desire to heal other's wounds, to get them back to a place of striving (Paton 2005; Paton et al. 2004; Sweeney 2012). This action of healing has nothing to do with getting a person back to what the leader sees as normal, but rather, getting them to transcend the hurt, and helping them become stronger by growing from it. When Spears (2004) identified empathy as a characteristic of the servant leader he wrote,

The servant-leader strives to understand and empathize with others. People need to be accepted and recognized for their special and unique spirits. One assumes the good intentions of coworkers and does not reject them as people, even if one finds it necessary to refuse to accept their behavior or performance (p. 9).

© Springer Nature Switzerland AG 2019

E. J. Russell, *In Command of Guardians: Executive Servant Leadership for the Community of Responders*, https://doi.org/10.1007/978-3-030-12493-9_6

This chapter focuses on the emergency services leader's responsibility to be an empathetic healer to the community of responders. As a leader, one is accountable for the health and wellbeing of those entrusted to your command. Emergency services cannot exist without emergency responders. One can take away the uniforms and apparatus; however, the emergency responder will still be the guardian on the hill. They are the essential component for effective emergency services response and are individuals willing to do whatever it takes to complete the job (Russell 2014).

Healing, like empathy, is not about getting somebody back to being "normal". It is about getting somebody to be whole again. A leader's responsibility involves being empathetic to followers and to self, realizing that all people need to be made whole again in times of trauma and loss. When Spears (2010) identified healing as a servant leadership characteristic he stated,

> One of the great strengths of servant-leadership is the potential for healing one's self and others. Many people have broken spirits and have suffered from a variety of emotional hurts. Although this is part of being human, servant-leaders recognize that they also have an opportunity to "help make whole" those with whom they come in contact (p. 9).

Empathy is a social approach towards people (Davis 1994). The empathetic healer can approach people as people; fully aware that they are in charge, yet not losing sight that the individuals in front of them are more than just their subordinates. It breaks down the armor that exists between people (Mortier et al. 2015). It changes the nature of human interaction. Being empathetic goes hand-in-hand with one's emotional intelligence (Burch et al. 2016; Humphrey 2015; Skinner and Spurgeon 2005). It transcends rank and strengthens the leader-follower relationship by allowing one to be human to one's followers. In the emergency services the rank is obvious: it is on one's collar for everybody to see. The empathetic leader, fully aware of their rank, can set it aside for the time being. The individuals standing in front of you know you are in charge; nothing has to remind them of that when they are in need (Norman 2016). One's position of authority is one of responsibility, but not monarchy. Allowing people to be themselves with you, getting them to open up to you, legitimizes your authority.

The emergency services leader as the empathetic healer realizes that their greatest asset is the responder. You can take away the rigs, the uniforms, the pomp-and-circumstance, and the responder will still be the responder. They will still be dedicated to the service of others. This is one of the great accountabilities for the emergency services leader. It is their responsibility to ensure that emergency responders can be made whole again. As discussed earlier, it is about the responder's posttraumatic growth. Over time, when faced with dramatic events and abnormal situations, one of two things can happen: the responder either grows stronger or they get weaker. The empathetic healer works to pursue the former. They walk with them through their pain, not to erase it, but rather, to put it into perspective. The weakening of the responder happens over time as traumatic events are allowed to fester and build upon one another. The empathetic healer knows that the end result of that scenario is burnout (Fishkin 1990, Paton et al. 2004; Sheehan and Van Hasselt

2003). For a community of responders, this is a dangerous place to be, especially when individuals are just trying to survive (Perez et al. 2010; Sweeney 2012).

The empathetic healer knows that individual situations will impact people differently. They are keenly aware of this and do not spend time trying to heal nonexistent wounds. As Thomas (2016) wrote, "A scenario or response may be insignificant to one individual but have a more significant impact on another (p. 58)". Knowing that the community of responders is made up of individuals, the servant leader will take the time needed for those individuals in need. That includes taking time for self.

Empathy for Self

If you stand at the American Cemetery in Normandy, France, a place where thousands of American soldiers and sailors are buried, this burden of leadership sobers you. Just being present at the entrance to the cemetery in view of that green garden of white ivory crosses and stars, cresting in the distance like a wave that will never break, is humbling to say the least. As one walks among the resting place of heroes, your consciousness reflects on the decisions some leaders must face. Somebody had to make that decision mindful of the fact that many were going to their death. And yet the leader also understood that they had to go; they were willing to give the order and live with the consequences. But imagine that crushing weight of such a decision.

The weight of leadership decisions within the emergency services can be debilitating. Emergency service leaders are the individuals who decide. Many seek the luxury of criticizing, but they have no idea of the devastating weight of a leader's decision. The final order comes from the leader. The ramifications fall directly on his or her shoulders. The burden is theirs. To go or not to go. To take a hill or not to take a hill. It is the leader, and the leader alone, that must live with that decision.

How does a leader reconcile the ramifications of their decisions? How does the leader come to terms with their orders? They do so by being empathetic to self. Society is captured in an instant news cycle full of relentless opinions. Far too often people don't realize just what it takes to send people into harm's way. When we look at tragedies and losses of firefighters, police offices, and EMS workers ordered to go, they did so under the command of a leader on scene that had to make that decision. The emergency services leader will ask others to do the impossible. To go even though self-preservation and human nature says otherwise. When things go wrong, that same leader will have to listen to the radio traffic recordings in the aftermath of an incident. That leader has to sit and hear a post-game recap of a team going to their demise. The leader will sit there and listen to the voices of the officer, the Mayday calls, and the alarms of pass devices and the calls for help over the radio. The tragedy will continue to play on repeat like a gruesome horror movie.

Imagine being that battalion chief of the Worcester, Massachusetts Fire Department the night of the Cold Storage Warehouse fire. This man had to physically put his body in the doorway to stop firefighters from entering the structure to continue to search for their already fallen comrades. Here is a man, dedicated to the job and under the crushing weight of a decision, who had to make the final decision that those inside could not be reached, that no one else could be sacrificed. That battalion chief's call was the right one; however, it was a crushing decision that will always carry a "what if".

This reality of emergency services leadership is why being an empathetic healer to one's self matters. Just like the responders, the leaders don't leave the scene unscathed. In order to be effective for the future, one must be empathetic to oneself and allow the wounds of leadership to heal.

Healing the Responder

The young and gung-ho responders are looking for opportunities to operate in high-stress situations. They want the orders, they want the calls, and they want to be the first through the door. They will enter into a situation like a bull in a china shop with little to no concern for their own safety. But it is the older, more experienced responders on scene, the seasoned leaders who must ensure that these individuals do not risk a lot to save a little. When you reflect back to a time when you were new to the job, you would internally question your officer's decisions to hold back. It's because you didn't understand at the time the ramifications of that which you were hungry for and desired. Yet now you find yourself as a leader, one that is responsible for the next generation of hungry responders, those same individuals that want to go.

The empathetic healer knows what it is like the first time a responder works a full arrest. Minutes seem like hours as they do compressions, establish an airway, shock the patient, set up an IV, and administer medications. All of that work, all of that adrenaline, all of that emotion, will seem in vain when the patient doesn't survive. That responder will spend the next 48–72 h going over that call in their head. They will question every decision he or she made. Like a film in their mind, they will try to remember every action, constantly wondering whether they did something wrong. The professional emergency responder, regardless of their time on the job, is a problem solver. If the patient doesn't make it, they didn't solve the problem. They will beat themselves up about it, and worse, if they do realize they missed something, or did something wrong that negatively affected the outcome, they will become wounded by it (Russell et al. 2015).

Healing the Organization

Like a tornado, poor and unethical leadership leaves a wake of damage. In order for the organization to move forward, that damage has to be repaired (Yancer 2012). The actions of poor leadership in the past can stifle trust now and in the future (Patterson 2003; Sipe and Frick 2009). Unethical leaders can create seemingly impenetrable walls of mistrust. Though you were not the leader or leaders responsible for what happened, you are responsible for what happens in the future. The empathetic healer that takes the reins of an organization understands that damage cannot be left unrepaired.

The wounds of poor leadership, hostile work environments, bully management, and unethical behavior will grow unless untreated. If an organization suffered at the hands of a poor leader, the newly appointed servant leader assuming command of that organization is completely aware that they have to heal the damage created by those who came before (Spears and Lawrence 2004). The servant leader is aware that the organization cannot move forward simply because they are now in charge. The concept of "there's a new sheriff in town" doesn't work. The wounds are there; an individual or individuals that sat in the seat that you now occupy created them. By occupying that seat, you will be viewed with mistrust by your followers until those wounds can be healed.

This isn't the fault of the community of responders. When individuals are abused and suffer at the hands of superiors, they cannot be held accountable for mistrusting people in authority (Undung and de Guzman 2009). The empathetic healer will not dismiss the damage that their predecessors may have caused. In fact, it is just the opposite. The servant leader, being the empathetic leader, will spend time with his or her followers, listening to them in order to try to identify the sources of pain (Gearhart and Bodie 2011; Reinke 2004). The servant leader will then strive to heal that pain and regain the trust of followers so that the organization can move forward.

It starts with empathetic listening and being open to hearing the truth (Frick 2011; Lloyd et al. 2015; Walker 1997). Success begins with a willingness to heal that damage. The servant leader understands that this takes time. Because individuals were wronged, they won't be eager and willing to share their thoughts and feelings for fear of reprisal. The servant leader knows that they have to work to gain the trust of followers (Patterson 2003). Followers must be able to believe that a leader's intent is nothing less than pure. It is here that the leader will fail if they even attempt to fake credibility (Greenleaf 1977/2002). If they pretend to care, that means that they don't care and sooner or later responders will become wise to the fact.

The empathetic healer also knows that the solutions have to come from the responders. Those who have been wronged have to be the ones that ultimately right the wrongs (Myers 2000; Welch 2015). The community has to have both a seat and say at the table. They need to be viewed as professionals and not victims. They need to be appreciated and listened to. The servant leader knows that until followers believe that their voices are being heard and their concerns are being addressed, trust can never take hold and healing cannot begin (Walker 1997).

Summary

Your actions and approaches as a leader become the learned behaviors of your followers. Your followers will be more likely to show empathy towards others if you are an empathetic healer as a leader. Behavioral research has found that it is quite common for individuals to mimic both the good and bad behavior of others. If their leader is an empathetic healer they themselves will be empathetic to their coworkers and those in the community in which they serve.

The empathetic leader, believing that it is their responsibility to help others heal, is a leader of great character. In fact, an empathetic healer is an individual of strong conviction, one who desires the best for their people and their organization. They take on their role fully aware that untreated wounds get worse. They understand that unless the wounds can be healed, the community of responders becomes vulnerable. Greatness comes from suffering that is healed, not suffering left untreated.

Empathy is about seeing people for who they are: people. Individuals come to the table with all of their greatness and all of their weaknesses, all of their abilities and all of their faults. There is no such thing as perfect, nor does there exist a "normal". Being empathetic doesn't mean that you seek perfection. It means that you recognize that your people can have problems and still do great things. People can be unique, yet still come together to bond as a team. In the end, because they are people, they deserve respect. A leader fails when they demand perfection. Organizations suffer when they try to define a specific best practice as it relates to humans. What makes people great are their series of life experiences coupled with their diversity. The servant leader recognizes this and uses individual diversity to the advantage of the organization.

Responders are nothing more than human beings that have a desire to serve. They bring with them all of the baggage that comes from being human. One's life experience off-duty shows up to work with the responder. If the individual is going through a family tragedy, the responder is going through a family tragedy. If the individual is having a substance abuse problem, the responder is having a substance abuse problem. They are one and the same, and no matter how hard one tries, they cannot be separated. The empathetic healer's role is simply to serve the individual who just so happens to be a responder.

Case Study A

As the chief of department, one of your battalion chiefs has just informed you that one of your most decorated and respected lieutenants was found intoxicated in his dorm room. The battalion chief also informed you that the lieutenant's wife just filed for divorce. This lieutenant is held in such high regards by his followers, peers, and senior leadership. You personally have known this lieutenant for 10 years. You were there when he was hired, and you were the one who tacked his bars on when he was promoted. Well aware of your department's substance abuse policy, you know this could mean the end of his career.

Case Study A Questions

1. How would you handle this situation?
2. What systems have broken down within this department?
3. Regardless of what your decisions are, what are the ramifications?

Case Study B

You are the Commander of a metro S.W.A.T. unit. Your team is called upon to respond to a home where a domestic violence situation has turned into a hostage situation. The situation involves a male subject that overpowered the first-on-scene officer, critically wounding him. The subject is now holding his girlfriend, her mother, and three children at gunpoint. After 7 h of negotiations, the subject becomes agitated and begins to pistol-whip his girlfriend. You then give the order to breech and enter the structure and neutralize the threat.

As the team enters using a flash-bang, gunfire is exchanged between one of your officers and the subject. When the officer, who was justified in using lethal force, returns fire, one of his rounds goes through the soft-tissue of the subject and into the head of one of the children, killing her instantly.

Case Study B Questions

1. How can this event impact your officers?
2. As commander, what is the impact of your decision to deploy your team?
3. What steps will you take going forward?

References

Burch, G. F., Bennett, A. A., Humphrey, R. H., Batchelor, J. H., & Cairo, A. H. (2016). Unraveling the complexities of empathy research: A multi-level model of empathy in organizations. *Emotions and Organizational Governance, 12*, 169–189.

Davis, M. H. (1994). *Empathy: A social psychological approach*. Winter Park: Westview Press.

Fishkin, G. (1990). *Firefighter and paramedic burnout*. Los Angeles: Legal Books Distributing.

Frick, D. (2011). *Greenleaf and servant-leader listening*. Westfield: Greenleaf Center.

Gearhart, C. C., & Bodie, G. D. (2011). Active-empathic listening as a general social skill: Evidence from bivariate and canonical correlations. *Communication Reports, 24*(2), 86–98.

Greenleaf, R. (1977/2002). *Servant-leadership: A journey into the nature of legitimate power and greatness*. Mahwah: Paulist Press.

Humphrey, R. (2015). The influence of leader emotional intelligence on employees' job satisfaction: A meta-analysis. In International Leadership Association 17th Annual Global Conference.

Katz, R. L. (1963). *Empathy: Its nature and uses*. London: Free Press of Glencoe.

Lloyd, K. J., Boer, D., Keller, J. W., & Voelpel, S. (2015). Is my boss really listening to me? The impact of perceived supervisor listening on emotional exhaustion, turnover intention, and organizational citizenship behavior. *Journal of Business Ethics, 130*(3), 509–524.

Mortier, A. V., Vlerick, P., & Clays, E. (2015). Authentic leadership and thriving among nurses: The mediating role of empathy. *Journal of Nursing Management, 24*(3), 357–365.

Myers, S. (2000). Empathic listening: Reports on the experience of being heard. *Journal of Humanistic Psychology, 40*(2), 148–173.

Norman, Z. (2016). Discovering my human nature as a servant leader. Available at SSRN 2773122.

Paton, D. (2005). Posttraumatic growth in protective services professional: Individual, cognitive and organizational influences. *Traumatology, 11*, 335–346.

Paton, D., Violanti, J., Dunning, C., & Smith, L. M. (2004). *Managing traumatic stress risk: A proactive approach*. Springfield: Charles C. Thomas.

Patterson, K. (2003). Servant leadership: A theoretical model (Doctoral Dissertation). Available from ProQuest Dissertation and Theses Database (UMI No. 3082719).

Perez, L. M., Jones, J., Englert, D. R., & Sachau, D. (2010). Secondary traumatic stress and burn-out among law enforcement investigators exposed to disturbing media images. *Journal of Police and Criminal Psychology, 25*(2), 113–124.

Reinke, S. J. (2004). Service before self: Towards a theory of servant-leadership. *Global Virtue Ethics Review, 5*(3), 30–57.

Russell, E. (2014). *The desire to serve: Servant leadership for fire and emergency services.* Westfield: Robert K. Greenleaf Center for Servant Leadership.

Russell, E., Broomé, R., & Prince, R. (2015). Discovering the servant in fire and emergency services leaders. *Servant Leadership: Theory & Practice, 2*(2), 57–75.

Sheehan, D. C., & Van Hasselt, V. B. (2003). Identifying law enforcement stress reactions early. *FBI Law Enforcement Bulletin, 72*(9), 12–19.

Sipe, J., & Frick, D. (2009). *Seven pillars of servant leadership: Practicing the wisdom of leading by serving.* Mahwah: Paulist Press.

Skinner, C., & Spurgeon, P. (2005). Valuing empathy and emotional intelligence in health leadership: A study of empathy, leadership behavior and outcome effectiveness. *Health Services Management Research, 18*(1), 1–12.

Spears, L. C. (2004). Practicing servant-leadership. *Leader to Leader, 2004*(34), 7–11.

Spears, L. (2010). Servant leadership and Robert K. Greenleaf's legacy. In K. Patterson & D. van Dierendonck (Eds.), *Servant leadership: Developments in theory and research* (pp. 11–24). New York: Palgrave Macmillan.

Spears, L., & Lawrence, M. (2004). *Practicing servant-leadership: Succeeding through trust, bravery, and forgiveness* (pp. 7–11). San Francisco: Jossey-Bass.

Sweeney, P. (2012). When serving becomes surviving: PTSD and suicide in the fire service. Retrieved from http://sweeneyalliance.org.

Thomas, C. (2016). Responder welfare. *Crisis Response Journal, 4*(11), 58–59.

Undung, Y., & de Guzman, A. (2009). Understanding the elements of empathy as a component of care-driven leadership. *Journal of Leadership Studies, 3*(1), 19–28.

Walker, K. L. (1997). Do you ever listen: Discovering the theoretical underpinnings of empathic listening. *International Journal of Listening, 11*(1), 127–137.

Welch, S. (2015). Leading from within: A modern view on leadership. *Griffin Leadership*, 31–34.

Yancer, D. A. (2012). Betrayed trust: Healing a broken hospital through servant leadership. *Nursing Administration Quarterly, 36*(1), 63–80.

Chapter 7
A Servant Visionary for the Emergency Services

Leadership is the capacity to translate vision into reality.

—*Warren Bennis*

Abstract This chapter introduces the visionary responsibility of the emergency services leader. A leader's vision becomes the map organizations follow in order to navigate a path towards success. The servant leader holds vision as a virtue, and the characteristics that form the servant leader strengthen their visionary ability. Yet, vision is only the first step. To bring that vision to fruition followers have to be the ones that work towards its reality. The visionary leader is the leader that dreams. They are keenly aware of the world, can conceptualize ideas, and have the foresight to see what does not yet exist. The servant leader is one who can persuade others to want to take this visionary journey and willingly make it their own.

Certain buzzwords seem to be overused in the study of leadership and management, and one of these words is vision. Textbooks and lectures on leadership hammer away at the concept of vision. Yet these works often fail to get to the heart of what a vision is and how one can work towards seeing it become a reality. First of all, a vision is no more than a conceptual idea. It comes from a leader's creative conscious mind (Fisher 2004). Unless acted upon, a vision is nothing; it is a thought at best. On the other hand, vision can be a check box for organizations that still spend time using strategic planning. Here, a handpicked group sits around trying to come up with a vision statement that captures the organization in the now, a statement that most followers within an organization aren't even aware of. A vision, and for that matter a vision statement, is meaningless if it only exists to take up space on a piece of paper or fill a bureaucratic square. For a vision to mean something it has to become something. One's vision should chart the organization's course; it is a roadmap to the future.

This chapter discusses the role of the serving visionary. It delves into the characteristics and constructs of the servant leadership philosophy, shedding light on how these attributes work together to bring a leader's vision to life. Instead of spending

© Springer Nature Switzerland AG 2019
E. J. Russell, *In Command of Guardians: Executive Servant Leadership for the Community of Responders*, https://doi.org/10.1007/978-3-030-12493-9_7

time simply selling the importance of vision, this section identifies how the emergency services leader can become a successful visionary by cultivating specific qualities within one's self (Russell 2014).

The visionary leader is an individual who is consciously aware, one who spends time in deep thought and learning. The concept of the vision is so much more than a simple statement. It is the fuel for forward motion. It is the coordinates that an organization will chart. It is an agreed upon ideal that the whole will work towards. The philosophy of servant leadership was a vision, one that imagined organizations working towards the good of people, society, as well as organizational sustainability and growth (DiStefano 1988). The vision of servant leadership would have remained nothing more than a theoretical concept if others were not persuaded to move the philosophy forward within academia and organizations. Researchers and corporate leaders who placed trust in the philosophy brought the vision of servant leadership to life. The philosophy of servant leadership does not sit on the shelf as a passing idea; it is not a vision statement. Instead, it is the driving force behind so many successful organizations. Today, servant leadership is Greenleaf's (1977/2002) vision transcended into reality.

For those in command of guardians, it is your responsibility to envision a future in which your organization can work towards. The status quo doesn't work. Most communities cannot afford a "one trick pony". The future is an all-hazards approach. If we look at where the emergency services are today compared to just several decades ago, so much of the career field has changed. Unfortunately, so much of that change was forced.

To be successful as a leader one must have both the ability to vision as well as the ability to persuade others to accept the vision (Spears 2010). Moreover, one has to have a vision that followers can believe in and desire to work towards. Followers are the ones that have to carry out the leader's vision. One's role as a leader transcends from a simple though to one of serving their followers so that they can bring the vision to fruition (Greenleaf 1977/2002). This entire process begins with awareness (Spears 2010).

A Leader's Awareness

Greenleaf (1977/2002) stated that the servant leader is one who is aware. One's awareness comes from one's consciousness within the world. It involves an awareness of surroundings, environment, opportunities, and threats. Essentially, it is being conscious of what's going on. Awareness is paying attention. It is mentally stepping outside of one's personal space to be conscience of occurrences and needs (Spears 2010).

Unlike one's civilian counterparts, the emergency services leader has had so many experiences in their career that awareness seems to come easy. The emergency services leader can look at a situation and know what needs to happen next (Strater et al. 2001). This awareness has been honed throughout one's career. Learning through both past occurrences as well as academic study forms the awareness of an emergency services leader. So much of the leader's ability to be aware is overlooked as a skill. It is dismissed as simply being "good at their job". The successful emergency services leader that is keenly aware has cultivated that awareness unconsciously. Most leaders haven't practiced awareness, nor did they ever set forth to become aware, so they don't realize the skill that they have. Their experiences are part of the job. Their education was done for both professional development and personal growth. In doing so, the leader cultivated their awareness (Jensen 2011).

The awareness of the leader also involves their ability to be self-aware (Spears 2010). It involves being truthful to one's self, honestly acknowledging one's abilities and weaknesses (Jensen 2011). Being self-aware of one's truths and abilities is a humble act (Patterson 2003). Such awareness assists the leader in the recruitment of others, individuals that may be stronger in areas than the leader. That self-awareness allows the leader to be conscious of their limitations. It is not in any way an admission of weakness, but instead, a virtue that leads to building powerful teams (Jensen 2011). Being self-aware opens the leader up to hearing ideas and thoughts from others and fostering a healthy environment of inclusive thought and open dialogue.

The advantage to the aware leader is their consciousness of both need and next. The aware leader is awake to the needs of individuals and situations and of the probabilities of what will happen next (Brindley and Tse 2016). Not through a crystal ball—but rather, through a cultivated mind that is open to information. The aware leader sees things in their whole. They don't fall into the trappings of tunnel vision, nor do they get stuck focusing on a single aspect of a situation. Again, the success of the visionary leader begins with their awareness (Greenleaf 1977/2002).

Conceptualizing a Vision

One of the attributes that separate humans from other animals is the human's ability to view their world in their mind—mentally conceptualizing possibilities and new ideas. When science says that elephants mourn, what they are saying is that they are capable of feeling sadness. But the difference between the human and the elephant is that the human is aware of their own mortality and can reflect upon it in the death of another. That conscious conceptual mind is one that can imagine what can be. The thoughts of a conscious leader can go beyond the known and imagine something that doesn't yet exist. Most people possess a basic ability to conceptualize. For example, they can imagine what a green room would look like if it were painted red. The servant leader can go so much further than that by imagining the possibilities of what his or her organization can achieve.

Unlike awareness, one's ability to conceptualize needs to be cultivated. The servant leader is willing to spend time in their own genius, to challenge their imagination and creativity, and to consciously look at the world as it is and imagine what it could be. Spears (2010) wrote that when it comes to conceptualization the servant leader,

Seeks to nurture their abilities to dream great dreams. The ability to look at a problem or an organization from a conceptualizing perspective means that one must think beyond day-to-day realities. For many leaders, this is a characteristic that requires discipline and practice. The traditional leader is consumed by the need to achieve short-term operational goals. The leader who wishes to also be a servant leader must stretch his or her thinking to encompass broader-based conceptual thinking (p. 22).

For the emergency services leader it involves thinking about what's possible. It is looking at the community of responders today and imagining the capabilities of tomorrow. It is conceptualizing what the future holds for the profession by imagining what will be its future responsibilities and what community needs will exist in the future. The servant leader within the emergency services can step back and imagine. They can envision a world where the knowledge, skills, and abilities of emergency responders are that much different than what they are today. It is dreaming of an organization that can deliver world-class service in a whole new way.

A great example of this is emergency medical services. The sheer cost and taxation of healthcare within a society will eventually drive changes associated with emergency medical service delivery. That means moving from standard street level treatment and transportation, to street-level diagnosis and intervention. To conceptualize what the paramedic will be in the future means that the leader has to envision a professional that is a practitioner of advanced skill and degree. To be successful at delivering emergency medical services in the future, the successful emergency services leader must conceptualize response capabilities and operations, which don't yet exist.

It is this conceptual dream that foster's a leader's vision (Spears 2010). This vision may be so big that it is possible the leader dismisses it as something unachievable. The emergency services leader cannot be afraid to take those big steps. To be successful, the leader has to be vulnerable with their ideas. Again, referring to emergency medical services, the current state of service delivery began with a conceptualized idea of what should be; it replaced an old model of response just decades ago. Additionally, it must be understood that when a leader conceptualizes and seeds a grand idea today, it is quite possible that he or she won't even be around to see it comes to fruition.

The Importance of Foresight

Greenleaf (1977/2002) argued that the leader who lacks foresight is a failed leader from the beginning. This has to do with the leader's inability to see what's coming or what may be needed. Foresight comes from a leader's understanding of the past, the present, and the possibilities of the future (Spears 2010). It is not some psychic ability, but rather, an educated guess based upon both the known and the possible. For instance, a police officer who puts a life flight helicopter on standby when responding to an auto pedestrian accident involving a child. A leader, based upon past experiences, as well as an awareness of what will be needed upon arrival, has the foresight to put things into motion.

As it is with conceptualization, a leader's foresight is a learned characteristic. It only comes from a leader's willingness to understand concepts, history, and options. It is a desire to be aware of the "what if's". Foresight comes from being awake to the world and open to knowledge (Spears 2010). It understands that in particular situations, certain things just happen. Foresight is the complete opposite of blind ignorance. It stems from the leader not wanting to be surprised. In fact, foresight allows leaders in the emergency services to be available for the unknown because they themselves already took care of the known (Marton and Booth 1997).

For example, picture an assistant fire chief of operations responding to a structure fire in a rural part of a county in the summertime. Part of that chief's success will involve responding extra assets for possible wildland and water supply operations while enroute. The chief does this because of past experiences. If that structure fire just so happens to spread to another structure, this would be the unknown. That chief's foresight already responded extra companies and equipment. Even though it is not being used for its original intent, at least it is there for that unknown spread. The chief's foresight allowed decisions to be made in real-time and not after-the-fact, when the consequences of time-pressure become unbearable.

Leaders who hone their foresight ability have strong convictions of not going against their gut. They understand that their training, education, and experiences subconsciously form their decisions (Jensen 2011). They have seen such scenarios and needs before. Maybe not as a carbon copy, but close enough that they mentally understand what the next move needs to be.

Again, like conceptualization, foresight is a learned skill. It can only come from a leader who desires to be aware. It is cultivated over time and plays a vital role in shaping one's vision. Without foresight, the leader is left battling issues and unknowns that should have been taken care of if only he or she were open to understanding the possibilities.

Persuading Others to Follow

Persuasion is the carrot in the carrot-versus-stick scenario (Gostick and Elton 2007). The ability to persuade involves getting people to see your way of thinking. It is the act of selling people on an idea, or in this case, a vision (Grenny et al. 2008). The power of persuasion is its ability to sway opinions and support in a non-coercive, non-bullying way.

Persuasion is grounded in trust. One's ability to persuade begins with their trustworthiness. Individuals willingly support your vision when they trust you (Caldwell and Hayes 2007). Followers will set forth to make your vision a reality when they trust you. This has to do with followers believing that you have their best interest, as well as the best interest of the organization, in mind. They trust that your intelligence and your understanding of the career field enable you to make solid decisions. Leaders fail at their role of persuading followers to carry forth their vision when they haven't taken the time to establish trust (Caldwell et al. 2009). This text will go in-depth on the construct of trust in a later chapter.

The servant leadership philosophy is about leading without force (Greenleaf 1977/2002). That's exactly what persuading people into wanting to be a part of a vision is about. It involves getting them to be willing to do the visionary work, and not because they have to or because they were coerced into it. A leader's success is realized when one can persuade others to support his or her ideas. When followers support the idea and believe in the vision, they personally desire to see it come to life. They view the vision as their future, supporting it as if it's their own. This belief in the leader's vision evolves into followers having a shared desire for the future and a willingness to work towards it because they want to be a part of that future.

Serving the Vision

The notion of serving one's vision is actually being in service to those that will carry out your vision. A great coach can envision a pathway to victory; however, it is the players that will actually do the work. The successful coach understands that in order for the players to be successful, their needs must be served. They need equipment, strength and conditioning, training, nutrition, and don't forget, contracts. It is not a TV movie. One hit without the proper padding, one shoulder to the jaw without a mouth guard and the coaching staff will be carrying the players off the field one-by-one. If these needs are not met, then the players cannot succeed. Meeting these needs is essentially serving the vision.

Serving the vision ensures one's followers can be successful in their pursuits (Fisher 2004). It circles back to an awareness of what their needs are. A willingness to serve the vision comes from an understanding that unless the basic needs of the follower are met, you as a leader cannot expect greatness. If your followers are still trying to get the basic necessities, using duct tape and bailing wire to hold it together, greatness is not in the cards.

Summary

Being in command of guardians, means being a servant visionary to the community of responders. It demands that the emergency services leader be keenly aware and awake to the world. It involves the leader being conscious and somewhat disturbed with the problems of the status quo (Greenleaf 1977/2002).

The servant visionary is one who hones their ability to conceptualize for a brighter future. It is about not being afraid to be open about, and address, the unknowns of tomorrow. It is about foresight and not ignoring what you know needs to happen. The servant leader knows that in order to be a successful visionary, they have to be able to persuade followers to accept the vision as their own. This starts with trust. Trust grounds one's vision in truth. Then it is all about serving the vision, and that means serving your people—the individuals that will bring your vision to life. For as de Pree (2011) wrote, "The first responsibility of a leader is to define reality. The last is to say thank you. In between, the leader is a servant" (p. 11).

As noted in the beginning of this chapter, vision is a concept that is thrown around but rarely analyzed. The servant leader knows that vision is not success, but rather, simply the first step. The fact is vision is just a by-product of a conscious serving visionary that is awake in the world, and fully aware that their conceptualization of an idea is only the beginning.

Case Study A

You have just been appointed the commissioner of a large fire and emergency services operation. Sitting on the desk in front of you is a report the last commissioner requested from an outside consulting agency. The biggest finding in the report is the cost-impact of emergency medical services operations. Your organization has realized a 7% year-over-year increase in the number of emergency medical service responses within the last decade. The report noted that this increase has dramatically impacted the organization, both financially and in its service availability. The majority of those additional responses involve populations that find themselves on the lower rungs of the socioeconomic ladder. The consulting firm discovered that the vast amount of patients were repeat customers known as "frequent flyers". These individuals rely on the emergency medical system for medical care.

Your emergency medical services division is the only area of the organization that can bill insurance companies for services rendered. The organization's apparatus, certifications, and training, as well as partial salaries for paramedics, are paid by these funds. Because this population lacks basic health insurance, your organization cannot recoup its costs. However, you belong to a public entity and have made an oath to serve, which means that care cannot be denied.

Case Study A Questions

1. How did a lack of vision lead to this reality?
2. What do you envision services looking like in the future?
3. How would you begin to tackle this problem?

Case Study B

You are the Under-Sheriff of a large county law enforcement agency. The Sheriff has appointed you to lead a committee that is looking into policing needs of the county over the next 20 years. The committee is tasked with developing a growth plan to meet the county needs. A report from the State shows that your county will experience a 7% annual growth rate due to the rapid expansion of technology firms and national financial institutions moving into your area. Besides the growth and need for more officers, the average deputy has worked for the department for 13 years. The job market outlook for the county is positive, with the median private sector salary exceeding $75,000 in the next 5 years.

Case Study B Questions

1. What are some of the problems that your agency will face in the next 10 years?
2. What steps need to be put in place to ensure your agency will meet the needs of the county in the future?
3. What will be the politically unpopular aspects of the committee's report?

References

Brindley, P. G., & Tse, A. (2016). *Situational awareness and human performance in trauma*. In *Trauma team dynamics* (pp. 27–31). Springer.

Caldwell, C., & Hayes, L. (2007). Leadership, trustworthiness, and the mediating lens. *The Journal of Management Development, 26*(3), 261–274.

Caldwell, C., Davis, B., & Devine, J. (2009). Trust, faith, and betrayal: Insights from management for the wise believer. *Journal of Business Ethics, 89*(1), 103–114.

de Pree, M. (2011). *Leadership is an art*. New York: Random House.

DiStefano, J. (1988). *Tracing the vision and impact of Robert K. Greenleaf*. Indianapolis: Greenleaf Center for Servant Leadership.

Fisher, J. (2004). Servant leadership it is the vision to see, and ability to serve. *Executive Excellence*, 15–16.

Gostick, A., & Elton, C. (2007). *The carrot principle: How the best managers use recognition to engage their people, retain talent, and accelerate performance*. New York: Simon and Schuster.

Greenleaf, R. (1977/2002). *Servant-leadership: A journey into the nature of legitimate power and greatness*. Mahwah: Paulist Press.

Grenny, J., Patterson, K., Maxfield, D., McMillan, R., & Switzler, A. (2008). *Influencer: The power to change anything*. New York: McGraw-Hill.

Jensen, M. (2011). Nurturing self-knowledge: The impact of a leadership development program. *OD Practitioner, 43*(3), 30–35.

Marton, F., & Booth, S. A. (1997). *Learning and awareness*. London: Psychology Press.

Patterson, K. (2003). Servant leadership: A theoretical model (Doctoral Dissertation). Available from ProQuest Dissertation and Theses Database (UMI No. 3082719).

Russell, E. (2014). *The desire to serve: Servant leadership for fire and emergency services*. Westfield: Robert K. Greenleaf Center for Servant Leadership.

Spears, L. (2010). Servant leadership and Robert K. Greenleaf's legacy. In K. Patterson & D. van Dierendonck (Eds.), *Servant leadership: Developments in theory and research* (pp. 11–24). New York: Palgrave Macmillan.

Strater, L. D., Endsley, M. R., Pleban, R. J., & Matthews, M. D. (2001). Measures of platoon leader situation awareness in virtual decision-making exercises (No. SATECH-00-17). TRW INC.

Chapter 8
Becoming an Emergency Services Leader for the Greater Good

*Never protect your past, never define yourself by a single
product, and always continue to steward for the long-term.
Keep moving towards the future.*

—Ginni Rometty

Abstract This chapter presents the concept of the emergency services leader becoming a leader who is a steward for the greater good. The steward is one willing to cultivate the organization and act as champion for its people and progress. They see the organization as a system, understanding that each of the system's parts has to work in order for the organization to be successful. The steward works for the greater good by rendering service to something bigger than self. They make meaning out of building a positive community of responders. They subscribe to the idea that followers have to be mentored so that one-day they are ready to be servant leaders themselves.

In his book *It is Your Ship*, Abrashoff (2007) wrote that when one assumes a leadership position over an organization it becomes "your ship". The achievement of this level is not the end-all accomplishment for the leader, but rather, the beginning of a new and daunting journey. Because it is your organization, you are ultimately responsible for it. You are accountable for everything that happens within it. As an executive level leader, when you look up from your desk, what do you see? How is your organization operating? What is the morale of your organization? What is the retention of your organization? As discussed in the chapter involving listening, what are the people, not just your inner circle, saying? These are questions that the keenly aware leader has to constantly be asking. The health of an organization, or more fittingly, the health of the people that make up the organization is vital. A failure of leadership is occupying the authoritative position over an organization and not being aware of what's going on within it. When you assume such a high level role of leadership you are agreeing to the responsibility and accountability of the organizational system, not simply the title, pay, say, or power that comes with it (Russell and Stone 2002).

© Springer Nature Switzerland AG 2019

E. J. Russell, *In Command of Guardians: Executive Servant Leadership
for the Community of Responders*, https://doi.org/10.1007/978-3-030-12493-9_8

Identified as a characteristic of the servant leader, one of the most accountable roles of a leader is that of stewardship (Spears 2010). The servant leader understands that one's role is not for self alone, but rather, for the greater good of the organization. As a steward, you are gifted the keys to the castle. This characteristic is so fitting for the emergency services leader because you don't own the organization—the taxpayer does. You, as the executive leader of this organization, are in fact a steward of the organization, entrusted by those who own it to watch over, cultivate, and care for it. You are responsible for ensuring that when the owner-citizenry is having one of the worst days of their lives and call for help, competent help arrives.

This again is one of those great differences between the emergency services and a private business; it is a reality that your charge is just temporary and one day you will move on. That means you cannot simply sell it off, close the doors and walk away, or move it somewhere else to benefit from a better tax structure. The organization that the emergency services leader is in charge of belongs to the people. You, as commander of an emergency services organization, are simply a caretaker of it for a finite amount of time.

The members that make up the organization are also temporary; their responsibilities are simply to be stewards to the people and to one another in their time of need. Again, unlike for-profit businesses, emergency services exist for one reason— the greater good in service to society. This concept doesn't mean that the emergency responder is in servitude to society, but rather, a steward to society. That relationship between leader-follower and follower-society forms what Donaldson and Davis (1991) called a "shared incumbency of roles" within the community of responders. As a steward-follower, one has both a responsibility to the citizenry they provide service and care to, as well as a responsibility to follow the orders and carry out the vision of those appointed above them. In return, the emergency services leader is responsible to society itself. It involves acting as steward of an organization that delivers services for the greater good in an exemplary way, all the while being a steward to the sum of the parts of the organization, specifically followers. This is a dual collective role that exists within the emergency services so that effective and efficient response can take place. Additionally, that effective and efficient response adds to the support of the end-user for the betterment of their organization meaning that when the taxpayer perceives his or her local response agency as one of high standards, professionalism, and competency, they will support that agency financially through the power of their vote.

One's role as steward begins with stepping back and looking up so you can see the organization for what it is: a complex entity with many moving parts, all of which are in need. Oftentimes, leadership misses this opportunity. Instead, they focus on the bureaucracy and place what matters to the side while concentrating on the mundane (Sergiovanni 2007). This chapter aims to draw attention to the concept of the emergency services leader as steward. It begins with a discussion of what it means to be an executive level steward for the greater good. It examines the concept of envisioning the organization as a system and not just individual entities that function on their own. This chapter presents how the leader can make meaning, gain trust from the public, and receive personal satisfaction from being a steward of a healthy community of responders. In addition, the chapter introduces the steward as mentor, a concept championed by Greenleaf (1977/2002).

The Greater Good

What is stewardship and what is the greater good? To illustrate this concept, we need to look at history. Centuries ago, stewards were trusted individuals that seemingly had the best interest of another in mind. They were gifted the power and authority of something that was not theirs. This fits well within the role of the emergency services leader due to the fact that he or she is entrusted with an organization that belongs to society. As Russell (2014) described,

> In medieval times, a steward was the "keeper of the hall" who had a mandate to administer the lord's estate in all matters large and small. The steward might delegate, but was personally accountable for everything that happened in the entire estate. Put another way, the steward was charged with the greater good of the estate (p. 38).

As for the greater good, that is the essence of what exists beyond self. The greater good is the wellbeing of others. The leader's service to the greater good is not altruism, but rather, benevolence. You, as the leader, are compensated for your work. You draw a salary and benefits for the level you hold. For the most part, your compensation takes into account the responsibilities associated with the position. This is the reason being a steward within the professional emergency services is not altruistic servitude; it is simply a successful leadership practice that makes life better for everyone.

Greenleaf (1977/2002) argued that a leader's focus on the greater good instead of self reduces one's irrational and selfish ego. This involves lifting up everything within the organization and the entities that the organization itself serves, to include the leader. It is here that the servant leader, in service to the greater good, personally benefits from said service. For example, most executive level emergency services leaders serve in their role at the pleasure of somebody else. Positions such as city mayor, city manager, a city council, a base commander, a commissioner, a governor, etc., all have power and authority over the executive level emergency services leader. When that executive level leader is a steward to the community of responders, the community of responders can operate and function at a higher level. That level delivers a quality product to the citizen and the public is satisfied. Because the public's needs are met, those that are above the executive level leader are also satisfied. This, in turn, strengthens the executive level leader's position by strengthening the trust they receive from those appointed above them. Additionally, delivering a successful public safety product can and often does equate to higher compensation and future opportunities above-and-beyond the position which they currently hold. Serving the greater good is not altruism; it is a benevolent act that benefits everyone.

Being a steward for the greater good within the emergency services ensures that you have the best interests of your followers in mind. As Block (2013) noted, being a steward of an organization means you consciously decide to refocus your purpose as a leader. Because you have been entrusted with an organization, it is up to you to cultivate that organization and to care for it. This means that approaching your role as a steward allows you to focus on the individual parts that come together to make up the whole. As a steward of the organization, you begin to believe in the power of empowerment and the delegation of authority (Block 2013). You take time with and get to know your people. You learn individual strengths and what areas need improvement. You do this for the greater good. You do this because you understand that when your people are empowered, when they are trusted and delegated to, loyalty to the organization goes up and the demand for greatness is realized throughout the follower ranks (Block 2013; Taft 2012).

The push back to being a steward for the greater good comes from the skepticism that some have. It is this idea that when we re-envision the approach towards leadership as a steward instead of an authoritarian, we are somehow asking leaders to relinquish power and become less (Block 2013). But this is not the case. As Greenleaf (1977/2002) noted, when you, as a leader, serve the greater good and transcend into being a steward of an organization, you achieve legitimate power. Your position is strengthened by each one of those individuals entrusted to you, who believe in, trust, and desire to follow you. It is not because they fear you, but rather, they have developed a love and loyalty to you for they know that you, as their steward, love them and are loyal to them (Taft 2012; Zohar 2000). Furthermore, stewardship considers the time of the leader. By empowering and delegating authority to others, it frees the leader up for self-healing (Lipsky 2009). The micro-manager and the authoritarian do not have time to take care of themselves because they are too busy trusting no one and taking care of business on their own. Being a steward of an organization is a healthy choice for a leader. Trusting your people and receiving that trust in return is healthy. Just look to the dagger in Julius Caesar's back to see what it is like to have people around you who do not trust you, do not love you, and want you gone. The stress and the daunting responsibility of being an emergency services leader are enough pressures for one to have. Therefore, having your people not trust and dislike you is completely unnecessary, and it causes an undue and avoidable burden.

Seeing the Organization as a System

Humans, things, and organizations are all systems made up of the sum of their parts. Each part cannot stand alone, for without the system as a whole the individual parts become irrelevant. Take a high-end sports car. Here you have an automobile of exquisite design, a mastery of engineering. However, as great as that vehicle is, you cannot remove its transmission and engine and place it in an economy car thinking that you are going to get the same performance that you do in the high-end auto.

Why? Because that high-performance vehicle is a machine constructed of high-performance parts that come together to form the whole. If you attempt to put a high-end transmission and engine into a basic car, even if it runs, it will not perform as intended. This is system thinking. It involves knowing that the organization is a makeup of its parts and that each of those parts cannot stand alone while others are having problems, glitches, or need troubleshooting (Jackson 2003).

The concept of system thinking has been around for almost a century. It is simply looking at an entity as a whole, realizing that in order to function at a high level, the whole entity needs to be properly functioning. System thinking is a way of approaching a problem or an issue with eyes wide open (Checkland 1981). Another way of explaining system thinking is looking at a lower back injury. Oftentimes, athletes and individuals who perform physically strenuous work injure or have severe pain in the lower back muscles. So, they stretch their back. They take anti-inflammatory medication. They try ice and heat. All the while ignoring the fact that the lower back muscles are attached to other areas of the body like the buttocks and hamstrings. The individual concentrates on the lower back trying to figure out why it is not letting go and feeling better, all the while ignoring the fact that they may have a knot in their buttock or a problem with a hamstring. They forget that their body is a system; they ignore the ways to actually solve the problems because they only focus on that one area.

To be successful as a steward of your organization, you need to concentrate on the entire system and not just an individual part. You could have procured some of the best equipment and vehicles that the emergency services has to offer, yet if your people are not trained and certified and thus given the opportunity to become the best they can be, that top-of-the-line equipment goes to waste. Emergency services organizations face this when it comes to technical and specialized operations. They are the sexy component of the emergency services. Responders want to be a part of a S.W.A.T. team or a technical rescue group, and of course, they want the best equipment. However, the equipment is only one part. The bulk costs and operations rest in the number of individuals you need who can perform that type of advanced operation coupled with a burden to get as many individuals as possible trained and certified in that area. And even at that, special operations are only one small part of the overall emergency services organization yet they consume large amounts of time and capital.

There are so many other operations, intricate details, and individual needs that must be served. It is system thinking that allows the leader to realize the need to be a steward of the organization and empower others. That allows one to give people the authority to make decisions and do the work. You, as a leader, are far better off finding individual strengths and talents in your responders that can take the reins of something and run with it. You are then free to serve that one empowered individual, who in turn will serve that particular program. Again, referring once again to technical and specialized operations: another individual might see a benefit of a multi-jurisdictional team that shares responsibility, staffing, and cost across agencies.

When an individual leader cultivates himself or herself as a steward and takes the time to understand system thinking, they seemingly abandon micro managerial habits and authoritarian approaches. The use of system thinking solves the leadership dilemma regarding what to let go (Jackson 2003). It is not about reducing people to processes, nor is it about getting out of one's responsibility as a leader (Checkland 1981). It is instead admitting that the organization is the sum of its parts and that there is no possible way that an executive level leader can focus on and control each individual piece. It is only through empowerment of others, and then serving those individuals, that the leader can truly succeed. It is about putting the system to work the way it is intended—one of skilled operations.

Making Meaning from a Healthy Community

As a leader, your success is measured by the success of the organization you command. Again, that organization is an entity consisting of individual parts. When you delegate authority and empower others, you are free to serve your peoples' needs (Greenleaf 1977/2002; Patterson 2003). As discussed earlier in the book, those individuals that are served and empowered in turn serve your needs as the leader (Russell 2016). It is here that you can make meaning for yourself as a leader and take pride in the fact that you have a successful organization.

As an executive level leader over an emergency services organization your greatest achievement is the healthy community of responders (Russell 2014). Think about what it will be like when you retire; are your people sad to see you go? When you look at the organization that you are in charge of, do you see community? The great irony of organizational assessment is that it is actually quite simple to measure. A successful emergency services organization will have strong retention, morale, and esprit de corps. There will be very little disciplinary actions. In addition, personnel issues, outside of individual needs such as benefits and pay, are rare. The steward ensures that the community of responders is healthy because that means the organization and its systems are healthy.

A Properly Served Public and a Healthy Emergency Services System

There is a direct correlation between a happy citizenry and a healthy emergency services organization (Bruegman 2012; Edwards 2010; Lloyd 2003). The healthy organization is free to deliver effective and efficient emergency services to the community. The steward recognizes that a healthy community of responders shows outwardly in relations and operations with the public. Issues of poor morale and unhealthy physical and dietary habits lead responders down pathways that are detrimental to achieving effective services (Lloyd 2003; Lipsky 2009). The greater good is that relationship between responder and citizen. It encompasses community relations, public education, as well as emergency response.

An unhealthy community of responders cannot cultivate a healthy relationship with those they serve, nor can they deliver superior services. The health of the community of responders goes hand-in-hand with emergency services delivery. The two cannot be separated. The steward must work to ensure that the responders are healthy so that the community can be served.

The Steward as Mentor

The final area for discussion is the steward as a mentor to others. Greenleaf (1977/2002) believed that the servant leader is a leader-mentor, that the role of the servant leader is to mentor others so they can grow. Greenleaf (1977/2002) asks, "Do they, while being served, become healthier, wiser, freer, more autonomous, more likely themselves to become servants" (p. 27)? The steward as mentor is the same as the servant leader as mentor. The goal is that those who are served and mentored by a servant leader one day become servant leaders and mentor others. Yes, as a servant leader, you can serve the needs of other people. However, if you desire (and if you are a servant leader you should) to have your people become servants to others, then you have to mentor them so that they can become servant leaders.

Earlier, when the text discussed empowerment and the delegation of authority, one of the reasons it matters so much is that if you, as a leader, are concentrating on the mundane bureaucracy, you are not taking the time to be a steward and a mentor to your followers. This is why the individual tasks have to be handed off. It frees you to mentor others. And as they grow, they themselves become mentors, stewards, and most of all, servant leaders. As a caretaker of the organization, one of your responsibilities is readying others to assume your role. Those who will succeed you must be ready to lead. Mentoring ensures one's followers are prepared to become future stewards.

Summary

This chapter addressed what it means to be a steward to an organization. The premise of stewardship involves the caretaking-cultivation of the organization as a whole. As an executive level leader of an emergency services organization, you have been entrusted with something that you do not own. You are its caretaker and its steward. The successful steward looks at an organization through a lens of system thinking and realizes that it is simply a sum of its parts. The steward recognizes that effective emergency services can only come from healthy responders. Furthermore, stewards assume the responsibility of building future servant leaders through mentoring, so that they can take command of the guardians long after one's tenure, thus allowing the organization to continue to serve the greater good.

Case Study A

One year has passed since you assumed your position as the deputy chief of operations for a large metropolitan fire department. One of the pillars of your responsibility is human resources—an area that includes personal actions and discipline. Since assuming your role, you have seen an uptick in personnel actions on individual members of the department. In fact, in the last 3 years, the last being under your tenure, the organization of 700 personnel has had to conduct investigations on an average of 6% of the department resulting in disciplinary actions 97% of the time. These actions range from minor infractions such as certification lapses and tardiness, to major issues like insubordination, drunk on duty, failure to meet annual physical fitness standards, assault, negligence, and patient abandonment.

These numbers are startling and the image of the organization is being tarnished. Before this trend and uptick in personnel actions, the organization would have to investigate less than 1% of its members annually, with needed disciplinary actions being a rare occurrence. In addition, for over two decades, the organization never faced a major personnel action against one of its members. In the last 5 years there have been major organizational changes to include the way the department promotes and staffs. Because of a local economic downturn, the now 700-member department was just 5 years ago an organization that consisted of 900 members. The current 700 members faced benefit cuts, mandatory overtime, and salary freezes. The current age of apparatus within your organization averages 8 years old. The last capital improvement made on a station was 6 years ago. To make matters worse, 39% of your force is eligible to retire.

Case Study A Questions

1. How would you, as the deputy chief of operations, approach this issue?
2. How would you communicate this finding?
3. What would you say is the impact on the organization? Not just from the personnel actions, but also, from the underlying causes that are driving these personal actions.

Case Study B

You have just been appointed the chief of police of a large metro police department. Last year, a male officer was involved in a physical altercation of a mentally challenged female subject wielding a baseball bat. In the altercation, the officer was injured and the female subject was killed at the hands of the officer. The incident was caught on a dash camera and the officer was both justified and exonerated by the district attorney. It needs noting that earlier that year, a robbery suspect armed with a large hunting knife was shot and killed by an officer that he lunged at and stabbed. After the second incident, a group of citizens demanded the officer be disciplined for what they deemed an overuse of physical force. Under immense social media induced public pressure, the police chief at the time suspended the acquitted officer and held a media briefing to explain that an internal affairs investigation had been ordered. Again, the officer was found to be justified and innocent of any wrongdoing; however, community relations are at an all-time low and so is the morale in the department.

Case Study B Questions

1. What is your role in this environment?
2. How would you go about being a steward to both the department and the community?
3. How do you see this event impacting the systems within your organization?

References

Abrashoff, D. (2007). *It is your ship: Management techniques from the best damn ship in the Navy.* New York: Grand Central Publishing.

Block, P. (2013). *Stewardship: Choosing service over self-interest.* San Francisco: Barrett-Koehler.

Bruegman, R. (2012). *Advanced fire administration.* Upper Saddle River: Pearson.

Checkland, P. (1981). Systems thinking, systems practice. In *Chichester.* West Sussex: Wiley.

Donaldson, L., & Davis, J. H. (1991). Stewardship theory or agency theory: CEO governance and shareholder returns. *Australian Journal of Management, 16*(1), 49–64.

Edwards, S. (2010). *Fire service personnel management.* Upper Saddle River: Pearson.

Greenleaf, R. (1977/2002). *Servant-leadership: A journey into the nature of legitimate power and greatness.* Mahwah: Paulist Press.

Jackson, M. C. (2003). Systems thinking: Creative holism for managers. In *Chichester.* West Sussex: Wiley.

Lipsky, L. (2009). *Trauma stewardship: An everyday guide to caring for self while caring for others.* San Francisco: Barrett-Koehler.

Lloyd, H. B. (2003). *Morale matters.* Memphis: Memphis Fire Department. Retrieved from www.usfa.fema.gov/pdf/efop/efo36355.pfd.

Patterson, K. (2003). Servant leadership: A theoretical model (Doctoral Dissertation). Available from ProQuest Dissertation and Theses Database. (UMI No. 3082719).

Russell, E. (2014). *The Desire to serve: Servant leadership for fire and emergency services.* Westfield: Robert K. Greenleaf Center for Servant Leadership.

Russell, E. (2016). Servant leadership's cycle of benefit. *Servant Leadership: Theory & Practice, 3*(1), 52–68.

Russell, R., & Stone, A. (2002). A review of servant leadership attributes: Developing a practical model. *Leadership & Organization Development Journal, 23*(3), 145–157.

Sergiovanni, T. (2007). *Rethinking leadership.* Thousand Oaks: Corwin Press.

Spears, L. (2010). Servant leadership and Robert K. Greenleaf's legacy. In K. Patterson & D. van Dierendonck (Eds.), *Servant leadership: Developments in theory and research* (pp. 11–24). New York: Palgrave Macmillan.

Taft, J. (2012). *Stewardship: Lessons learned from the lost culture of wall street.* Hoboken: Wiley.

Zohar, D. (2000). *Spiritual intelligence: The ultimate intelligence.* New York: Bloomsbury.

Chapter 9
Serving the Responder's Growth

Growth means change and change involves risk, stepping from the known to the unknown.

—*George Shinn*

Abstract This chapter discusses one of the greatest responsibilities of the emergency services leader: serving the growth of their followers. The willingness to serve the growth of followers is grounded in a desire to want to see followers change and grow, so that one day they too can become leaders. The servant leader serves the needs of their followers in a way that positively impacts their mental, physical, and emotional health. These three areas exist in unison; the successful leader realizes that they cannot stand alone. The healthy follower becomes less vulnerable, realizing posttraumatic growth. The emergency services leader that commits to the growth of their followers does so with an understanding that such growth reduces vulnerabilities associated with traumatic experiences and protects responders from burnout.

For so many leaders in the emergency services, their time is not spent on developing their people, nor is it spent in command of an incident. The bulk of one's time is spent doing superfluous work. On average, emergency service leaders find they spend a majority of their time sitting in meetings, answering emails, and doing paperwork. As discussed in earlier chapters, the emergency services leader came into the career field with the desire to serve others, but the pressures and cultural habits of the profession move one away from that desire and instead immerses them in a bureaucratic stranglehold.

Nobody's saying that all meetings are worthless, nor is a movement afoot to do away with all policies and procedures. Instead, balance is needed. Leaders bogged down with the daily mundane unintentionally impact their followers negatively. Followers need their leaders to be among them—not all the time, but enough to make a difference. That is, leaders must demand to have the time to spend with their people so they can nurture and help them grow.

E. J. Russell, *In Command of Guardians: Executive Servant Leadership for the Community of Responders*, https://doi.org/10.1007/978-3-030-12493-9_9

When Spears (2010) identified commitment to the growth of followers as a characteristic of the servant leader, he stated,

> Servant leaders believe that people have an intrinsic value beyond their tangible contributions as workers. As such, the servant leader is deeply committed to the growth of each and every individual within his or her organization. The servant leader recognizes the tremendous responsibility to do everything in his or her power to nurture the personal and professional growth of employees and colleagues. In practice, this can include (but is not limited to) concrete actions such as making funds available for personal and professional development, taking a personal interest in the ideas and suggestions from everyone, encouraging worker involvement in decision-making, and actively assisting laid-off employees to find other positions (p. 23).

Because you find yourself in command of guardians, you are responsible for cultivating the guardian's growth. That is what this chapter is about: leaders spending the time serving the growth of their responders.

This role of leadership as serving the follower's growth is one of the most essential responsibilities of the emergency services leader. The individual responders are the most important assets. The better they are, the stronger they are, the more competent they are–the greater the impact on emergency services delivery. The responder must be cultivated and developed. The recruit candidate academy and the basic academic education is just the beginning. When a responder enters the emergency services career field, they are stepping into a world where survival depends on one's ability to learn and grow. Additionally, to survive the profession, they must become stronger and more resilient.

Growth of the responder all comes down to understanding and working towards strengthening their mind, body, and spirit needs holistically. This is the reason why this book offers the philosophy of servant leadership to those in command of guardians. The foundational question of the philosophy of servant leadership asks, "Do those served grow as persons" (Greenleaf 1977/2002, p. 27)? It is the core question that drives everything else. It is the question that every leader within the emergency services must ask. This is no longer a good idea, but rather, an essential task. The health and wellbeing of emergency responders is directly correlated to their growth as persons (Henderson et al. 2016). In order to grow and preserve their humanity, their leaders must serve their needs.

This growth of individuals does not just occur in the early years of a follower's careers. It is ongoing and inclusive. That means that the individual executive level leader needs to continue to grow. Part of serving the community of responders is the leader serving his or her self as well. The executive level leader must continue to grow, they must continue to learn, and they must remain physically, mentally, and spiritually strong. This means taking time for self, cultivating one's mind, scheduling time for physical activity, finding a trusted confidant to talk to, and taking time away from work. This is basically the oxygen mask on an airplane; you have to put yours on first before you can help somebody else.

This chapter identifies different areas that leaders can concentrate on in order to foster follower's growth. It begins with a leadership exercise, and then moves on to talk about the importance of succession planning, as well as developing responder resiliency. It then continues on to discuss mental and physical needs, making the case for why and how leaders need to concentrate on these areas to increase the wellbeing of the community of responders. This is the essential role of the emergency services leader: serving their followers so they can grow and serve.

Leadership Exercise

A great exercise that one can do to glean an understanding of the type of leader they are is to reflect on what one spends most of their time doing, thus honestly acknowledging what gets the majority of one's focus. It is a simple exercise that takes less than an hour to do. One needs to have a pad in front of them so they can write down the answers. Understand that the hardest part about this exercise, if it is done honestly, is both coming to terms with the findings and being willing to change.

On a piece of paper, make four columns and 10 rows. Begin by listing the 10 things that seem to use up most of the time during the day (and remember, be honest). Note that emergency response is not a part of this; that is simply a given and the reason for the organization's existence. In the next column, reflect on each one individually, and then assign each of them their own priority, using 1–10. For example, a 1 would represent that this task occupies little to no difference in the lives of responders; these are the things that can wait, things that can be delegated, and no life is dependent on getting this accomplished. A 10 represents something that is essential to the responders. These are tasks that cannot be delegated, and must be accomplished. In the third column, using a few words, write down why it has the number assigned that it does. Then in the fourth column, write down how much time per week in minutes is spent on each of those tasks.

By doing this, it gives the leader a real snapshot of their week, showing what is truly being prioritized and consuming one's time. For many, the biggest shock is discovering what is taking the most time vs. what has a high priority. Only the leader can answer this. By putting it down on paper, it begins to make sense because it allows you to see it for what it is.

Succession Planning

In an earlier chapter, we discussed the servant visionary being one that is aware of and can conceptualize the future. They are the dreamers, the ones who can conceive the future in their mind. The servant visionary is a committed leader to both the growth of people as well as the future of the organization. They understand that one day they will no longer be in command of guardians. They are keenly aware that those who come after them will be asked to assume that role. The succession needs of the emergency services once again make the case for servant leadership within the emergency services career field (Keith 2008). For as Greenleaf (1977/2002) asked, "Do they, while being served, become healthier, wiser, freer, more autonomous, more likely themselves to become servants" (p. 27)? If a leader desires a servant leadership organization, they must cultivate an organization of servant leaders.

Part of being committed to the growth of individuals involves developing them into the leaders of tomorrow (Groves 2007; Rothwell 2010). Your followers are the ones that will fill the void after you are gone (Russell 2014). Succession planning starts with growing the individual responder both as a person and as a future leader. It does not start down the road; succession planning is successful when it starts today (Michelson 2006; Seigal 2006). The premise is getting one's followers ready for the responsibilities of tomorrow (Michelson 2006 Seigal 2006). It involves cultivating their desire to serve, that same desire to serve that you have, and nurturing it into their approach towards leadership.

As Rothwell (2010) noted, it is about constructing the leaders of tomorrow in your people today. Leaders come from within the emergency services, it is rare at best to see someone promoted to an executive level position that has not come up through the ranks (Russell 2014). Unlike other areas of government where executive level leaders may come from other industries or sectors, the emergency services require a rich understanding of the profession that only comes from serving as a responder. This is both a benefit and a problem within the emergency services. On one hand, individuals understand the roles and responsibilities associated with emergency response. Yet on the other hand, just because an individual can command an emergency scene, doesn't mean they are ready to be leaders away from the emergency scene. To be a leader, it takes education and training, coupled with experience and mentoring.

To be successful at succession planning, a leader must be aware of their ego and humbly give of themselves to develop individuals to take command of more than just the emergency scene once they are gone (Michelson 2006; Seigal 2006). The success of the servant leader, as it pertains to succession planning and programs, has to do with the servant leader not being concerned with job protection and positional preservation (Groves 2007). The servant leader believes that their success as a leader is measured in part by their follower's ability to seamlessly take over when they leave, not how long they occupied their position (Greenleaf 1977/2002).

One of the easiest and most effective ways of fostering emergency service leaders of tomorrow is to use empowerment beyond just emergency response, giving individuals responsibility, and allowing them to grow from those experiences (Patterson 2003). This allows followers to be accountable for decisions, programs, and projects. It cultivates their ability to plan, to think critically, and to make decisions. Empowerment is an area where the emergency services succeed. The profession operates on junior officers making on-scene command-and-control decisions without the direct supervision of senior leadership (Russell 2014). Moving that empowerment into non-emergency functions does not take much effort; however, it does take will.

The use of empowerment, plus education and mentoring, is the easiest and most effective way to answer the responsibility of succession planning (Groves 2007).

This is a best practice. It cultivates followers, readying them to take on the responsibilities of tomorrow. The key to succession planning is not putting off till tomorrow what can be done today (Michelson 2006; Seigal 2006). Waiting for the individual to be promoted into a position before they are cultivated as a leader hurts the organization by slowing, or worse stopping, forward progression in order for the newly appointed individual to catch up in their abilities.

Resiliency

One of the most grueling and disciplined athletic pursuits is that of a boxer. To become a professional boxer, an individual will spend years training. They will take part in some of the hardest physical exercises—developing endurance other athletes find themselves envious of. So much of the cardiovascular, strength, and conditioning routines are not done so they can throw a punch, but rather, so they can take a punch and come back swinging.

This is resiliency. It is the ability to take a hit and come back. To the emergency responder that hit can be physical or mental. This is why their growth is so important. The essential goal for the emergency services leader is to develop and grow his or her followers so they can find strength in their resiliency. The aim is the responder's posttraumatic growth which cultivates their ability to come back swinging (Paton 2005).

This concept has nothing to do with weakness, but rather reducing vulnerabilities (Rutter 1987). As it is with a fighter, so it is with the emergency responder. They can be great at their job and their skills and abilities can be second to none; however, if that individual responder is mentally or physically vulnerable, then they are not resilient to the traumas of the profession (Surjan et al. 2016).

Resiliency is not about removing the responder's humanity; that is what makes them human. Resiliency, as it relates to the emergency responder, is about strengthening them physically and mentally long before they face the traumas. That physical and mental strength allows them to take the traumatic hits and grow from them (Satici 2016). It in no way implies that you can get an individual responder to a point where they are not impacted from trauma. Instead, it is about getting them to a place where they can grow from that trauma.

Fostering mental resiliency is about putting the traumatic experience into perspective. It is reassuring the responder that the way that they feel is not abnormal, but rather, a natural reaction that can be healthily navigated (Rutter 1987). Physically, it is not about not getting hurt, but rather, it is about being able to come back healthy and strong when physically impacted (Surjan et al. 2016).

It needs to be made clear that no matter how great one's resiliency is, there are times when the responder just cannot bounce back. It is at that point that emergency services leaders need to recognize other opportunities for that individual, being consciously aware, open, and truthful to the fact that some experiences and injuries are just too much for an individual to handle (Paton 2005). It is at that point when the responder acknowledges they have done their duty, have served honorably, and now know it is time for them to move on to new opportunities (Mitani et al. 2006). The decision to move on is a noble one, one that emergency services leadership needs to support and assist. Going back to the growth of the individual, part of resiliency is encouraging them to learn skills that can be used outside of the emergency services profession. This goes to the heart of ensuring a responder's physiological and safety needs and reminding them that they do not have to remain in a profession they are no longer meant to be in just because they lack options.

The emergency services leader is accountable for this cultivation of resiliency in his or her followers. It is one of the greatest responsibilities that they are tasked with. Leaders need to know what areas to work on—areas that reveal results. Moving forward, let's delineate on these areas of focus where leaders can make a real difference in the resiliency of the responders, and it all starts with mental health.

Mental Health

One thing that society and self-help books have in common is this notion that an individual needs to feel happy all the time. There are prescription pills and conferences to help individuals try to find constant happiness. Today, it seems that if one is not happy all of the time then something must be wrong. This idea is a fallacy. It is simply not possible for an individual navigating the human experience to be happy 100% of the time.

Happiness needs to be viewed as a state of being, not an emotion (Suzuki and Fitzpatrick 2015). Yes, happiness relates to mental health; however, if an individual is feeling unhappy that does not mean they are mentally unhealthy. All it means is that they are human. This idea to force a constant state of happiness can weigh heavily on the mind of an emergency responder. Here is an individual that just encountered a horrible experience, now getting off duty and going home to his or her family. The family is left wondering why the responder is not happy when they get home. So that same responder who just experienced a trauma is now sitting there in front of their family, thinking that something is wrong because they do not want to play with the kids or talk about the upcoming house remodel. This situation creates an existential crisis for the responder; leaving them thinking that something is wrong with them (Kirschman 2004, 2006). The inevitable frustration from the family members does not help the situation either (Kirschman 2004, 2006).

The only way to combat this is to help the responder understand that they are not mentally unhealthy; they are simply reacting based upon the conditions of the experience they just had. Their mood is not permanent; it is just a temporary state that is part of both the grieving and healing process.

The idea of mental health of the emergency responder begins with strengthening their knowledge about the subject of mental health. It is making them aware that there will be times when they will find themselves in a psychologically dark place (Mitani et al. 2006). Their mental health, or mental resiliency, is the flashlight and guide rope that will help them find their way out of the darkness. Unlike what society is selling, it is fine to be in a dark place for a while. Individuals bounce back in a time that is right for them, not others around them.

The emergency services leader's responsibility is not being the psychologist for his or her followers. Instead, it is recognizing that there are areas in which you can strengthen mental resiliency of followers (Meyer et al. 2012). As a servant leader, being committed to the growth of your followers is an area you have power to focus on. It is not necessary to be a trained mental health worker in order to foster these areas within the community of responders. Instead, you as an emergency services leader, just have to be made aware of them, understanding that they go hand-in hand with the mental health of followers.

As a leader, the reason for this is simple; it is so followers today can continue to be followers and future leaders of tomorrow. It involves realizing that in the emergency services, burnout is a real threat (Mitani et al. 2006). The healthier responders are mentally, the better their chances for developing posttraumatic growth (Paton 2005).

Academic growth One of the easiest places for a leader to start is in the academic growth of followers. There is a direct correlation between mental resiliency and intelligence (Duckworth and Eskreis-Winkler 2013). As individuals grow in their knowledge and understanding, they strengthen their ability to think critically and make decisions. One's ability to think critically allows them to navigate the sometimes-unknown world of reaction to tragic experiences. The critical thinker is conscious of their mindset knowing that the way they feel is both natural and personal (Dweck 2006). The individual's ability to think shapes the way they navigate society. It allows them to put things into perspective.

Enriching one's ability to think is the cornerstone of the academic experience (Lopez and Louis 2009). Unlike anything else, academic growth relates directly to an individual's ability to set aside bias and be open to the facts. This is the intangible benefit of academic growth, one that changes and shapes an individual's worldview (Dweck 2006). The stronger an individual's ability to think is, the more resilient their mind is when faced with new and challenging experiences (Duckworth and Eskreis-Winkler 2013).

For the emergency services leader, the academic growth of followers can be something as simple as fighting for education benefits and time on duty that can be dedicated to coursework. It is championing academic growth; not just for promotion, but also, for feelings of personal achievement. It is being a leader that believes in learning just to learn, desiring a community of responders that is thinking and conscious. It should be noted that one thing great leaders throughout history have in common is that they are well read and open to knowledge.

Personal enlightenment From academic growth comes personal enlightenment. This is the change in mindset that Dweck (2006) identified. The concept of personal enlightenment is a consciousness of who one is, what one's abilities are, what one's potentials are, and the environment in which one navigates (Gallagher 2016). That state of consciousness allows the responder to be aware of their feelings and emotions (Newberg and Waldman 2016). The enlightened mind is one that can put things into perspective. It is transforming the individual's interpretation of their role and existence within the world (Newberg and Waldman 2016).

Personal enlightenment involves becoming aware of one's potential. This is vital for the emergency responder because responders have to see themselves as responders in order to truly become responders. They need to experience what that means; they have to do the job. Enlightenment happens once the individual responder transcends from the graduate of the academy to being part of the com- munity of responders and begins to do the work. Over time, they become enlightened to what it means to be a responder, as well as who they are in this role within society (Newberg and Waldman 2016). Furthermore, the enlightened responder becomes both aware and excepting of the human experience, of the notion that some live and some die, that life itself is a cycle.

Personal enlightenment leads to a responder's awareness that tragedy exists. That enlightenment does not make the trauma any less, it just allows the responder to know that all they can do is their best, try to save as many as they can, and that sometimes things are just out of their control.

The emergency services leader cultivates an individual's path towards personal enlightenment by promoting open dialogue about experiences and finding opportunities for growth. One way is recommending existing literature that helped you grow as a responder into the leader you are today. A great way to do this is to simply put them on the shelves in your office and invite people to borrow them. Another way is finding conferences and retreats that focus on personal growth, along with funds, to send your people to. These are realistic actions that the emergency services leader can take to help cultivate his or her responder's personal enlightenment. Again, it is a personal journey for the responder, one where the leader just needs to walk next to them.

Awareness of self For followers to one day become servant leaders, to follow in your footsteps as a servant leader within the emergency services, they have to become self-aware. The servant leader is also that servant visionary who is aware of self, the needs of others, and the environment. Fostering the emergency services responder's self-awareness means assisting responders on the pathway to knowing who they are, the needs of the populous they serve, and the world in which they navigate.

A responder's self-awareness correlates with their ability to empathize with both others and self (Hamm et al. 2016). This ability is important for forgiving self. There will be times where the responder will not be successful. No matter what they do, no matter how good the intervention, the outcome will be negative. That self-awareness allows them to know their limitations, that they are not perfect, and that things will go wrong even though they did everything right. It leads to their ability to forgive themselves when things go wrong.

An awareness of self leads to a healthy state of being (Salovey and Mayer 1990). It allows one to be confident of one's abilities and skills, yet aware that sometimes there is nothing one can do and accepting that as a truth. It involves learning that one is simply a person, and the best one can do is become aware of self, be open to change, and live in a state of personal growth (Marton and Booth 1997).

An awareness of self allows one to become conscious of one's boundaries. One who is self-aware is both confident and humble in their person. This plays a role in the responder's interactions with victims and their ability empathizes with those who suffer (Hamm et al. 2016). An awareness of self, cultivates one's ability to see the humanity in another and to not lose touch with the fact that those they serve are people (Hamm et al. 2016). It allows the responder to do his or her job without judgment or prejudice, because their self-awareness of their own weaknesses strengthens their humanity, kindness, and compassion towards others.

Frequently, responders that have been on the job for a while are referred to as "jaded". Self-awareness changes that. Instead of feeling "jaded", the self-aware responder is just conscious of the human condition. It allows them to grieve in a healthy way and to see occurrences as they are. Many seasoned responders can be perceived as "jaded"; however, they are simply professionals that understand that there is a time for emotion and a time for action.

Emergency services leaders should help responders resist becoming numb to the human condition. Leaders can help put things into perspective by helping responders reflect on the situation. That reflection aids with coming to terms with the outcomes. This demands that the leader be honest of their experiences and feelings, as well. They should be open to having their stories heard, letting responders know how they navigated it, and how their feelings and emotions were in the aftermath. It is about not hiding the real self, and being willing to share one's triumphs and losses. This adds to a leader's credibility; it allows followers to see one as a human, and not to fear their emotions. This practice overcomes the hyper-masculine fallacy.

Emotional maturity There is a direct link between one's emotional maturity and their career satisfaction (Humphrey 2015). In this case it is not about maturity as one who's always serious or silly. Emotional maturity is about developing one's mind to allow for healthy emotional navigation of situations and interactions (Salovey 1997). Part of emotional maturity is the concept of time and place. There is a time for seriousness and there is a time for silliness. Many responders will find themselves in situations that naturally stimulate some type of emotion response. In these instances, someone who is emotionally mature understands their emotions and how to manage them (Humphrey 2015). The responder that is emotionally mature understands that he or she is in control of self (Salovey 1997).

The emotionally mature individual is one that is in control of their future (Humphrey 2015). They are individuals that are self-aware of their wants and desires. The emotionally mature responder understands how they want to be perceived by others. It is reaching a level where the individual accepts the fact that they are the one in charge of their behavior.

The emotionally mature leader impacts the emotional maturity of one's followers (Salovey 1997). This is as if one's emotional maturity becomes a learned habit, one where followers learn how to behave based on their experiences with their leaders. Human beings become what they experience. If followers are subjected to an emotionally immature leader, those followers are more likely to become emotionally immature leaders themselves (Salovey 1997). Servant leaders that are aware of this cultivate their own emotional maturity so that they can personally influence their follower's emotional growth.

Work life balance It is that age-old question, do you live to work or do you work to live? There are those within organizations that can find themselves at a point where they begin to believe if they take time off the world will fall apart. For those in the emergency response career field, their role becomes a part of their identity. It's worth noting, however, that it is not the only part. Because of the nature of the career, emergency services work requires healthy time off. No matter how strong somebody thinks they are, the work and the bureaucracy will get to them. Healthy time off allows individuals to refresh and recharge (Mas-Machuca et al. 2016).

Healthy time off allows responders to put their work into perspective, allowing them to see that the world is not in a permanent state of suffering and tragedy. Yes, it is a noble profession and the work that they do is vital, but it is not all that they are. In order to healthily navigate the pressures of the job, responders need time off to see life as a thing of beauty and not just tragedy (Hougaard et al. 2016).

Far too often you hear stories about leaders and organizations that give their people a hard time about taking a vacation or a personal day. Followers within those organizations suffer. Even when they really need the day off for personal reasons, they fear being made to feel guilty, reprimanded, or badgered by management. Individuals that are involved in the wellbeing and protection of others are made to feel guilty for taking a day off when they themselves are ill. That is a learned bullying behavior that emergency services leaders must ensure does not go on in their organizations. It is a fallacy to believe that the more time people spend work the better they are at their jobs. Individual responders that take vacations and time off are healthier and more productive than those who do not (Russo et al. 2016).

In fact, the less time people have to enjoy their life and family, the less productive they are (Hougaard et al. 2016). Emergency services leaders have a responsibility to make sure that their people are getting healthy and much needed time off. They must be aware of followers who are not taking leave, and be willing to have an open dialogue with them about why. The more time an individual responder spends at work, the more at risk they are for burnout (Mitani et al. 2006).

To take it even further, emergency service leaders need to be aware of those companies that are overburdened with responses. For example, you may find yourself in charge of an agency that staffs 11 fulltime ambulances throughout a response area, and one of those ambulances, on a daily basis, sees double and sometimes triple the amount of calls more than the others do. You, as a leader, need to ensure that people are rotating in-and-out of that assignment, that they are being backed up by other companies, and that personnel assigned to those positions are taking leave. This scenario is common in so many emergency services organizations; there is always a specific area within a larger response region that will have a greater demand on the agency's resources than others.

Physical Health

There's a direct link between an individual's mental health and their physical health (Henderson et al. 2016). One's physical health is one's physical being. It is the way the individual physically experiences the world. One's physical health is usually the first impression that others have of the person. It is the way in which an individual presents themselves to the public. In order to promote the growth of responders so that they can realize their full potential, it is imperative that leaders understand the vital relationship between the mind and body.

Physical health of the emergency responder matters for many reasons. The first is that the physically healthy body strengthens the mind. Physical conditioning and exercise create positive feelings and emotions (Carey et al. 2011). They give the individual a sense of achievement as they grow in strength and endurance. Physical activity pushes human limitations, providing the mind with a sense of accomplishment. It constantly gives the individual something to work towards. It could be a goal of running a marathon, participating in a triathlon, taking part in powerlifting competitions, or whatever gives people ongoing goals to reach. Physical activity provides purpose. That purpose strengthens focus, one's ability to commit to something, and a hunger to be better tomorrow than what one is today.

It needs noting that the physically healthy body can withstand the impact of emergency services work (Carey et al. 2011). There is a myth in the emergency medical services that believes the more stairs and turns there are in the structure the larger the patient will be. Regardless if that is true, as our society grows in its physical stature, emergency responders will face more victim movements and physical altercations involving larger individuals. This, along with the impact of the work, the bulkiness of the gear, and the weight of equipment all takes its toll on the responder's physical structure. Over time, the physical stress on the body can start to wear it out. The physical health of the responder lessens the impact of the work by strengthening the structure itself.

Another reason to concentrate on physical health has to do with the mundane monotony that so many responders face in between calls. There is a lot of downtime, and that time needs to be filled with something. Championing physical fitness and healthy nutrition within the community of responders offsets the idea of filling the boredom with more lazy activities and unhealthy snacking. This is easily accomplished when leaders build physical fitness programs and nutrition-focused classes into the daily schedules (Hoeger and Hoeger 2016). Doing so adds routine by making mandatory physical fitness a part of the workday.

It is a leadership pitfall to set physical fitness programs aside as something to do when everything else is done, instead of making it a part of the culture. If physical fitness is not part of the day-to-day activities of the community of responders, then everything else will take precedence over it. Physical fitness programs must be as natural to emergency responders as checking out their apparatus and attending training classes (Hoeger and Hoeger 2016). It just has to be a part of who they are and what they do. It starts by seeing the emergency services responder for who he or she really is: an athlete. What society demands from emergency responders mimics athletic competition. No athlete can be successful if they are physically unhealthy and skip their training days.

To be a successful emergency services leader, one has to be a champion for the physical health of responders. By law, physicals are required, so that part is taken care of for the leader. Being physically healthy on the other hand is not required, yet ironically, it is an imperative need. The unhealthy responder cannot be a resilient responder. The unhealthy responder cannot grow into the successful leader. That's because the unhealthy responder will eventually fall. We know this to be true with emergency responders, because 50% of line-of-duty deaths involve heart attacks. It needs noting that some cardiovascular problems are genetic, yet even with genetics, physical activity and overall health greatly reduces the chances of developing problems. But the truth is so many emergency responders fall because of unhealthy living and lack of physical fitness. In order for responders to grow and one day become leaders, they have to survive being responders.

For the emergency services leader to truly serve the needs of their followers, they have to be honest with their followers. They have to set the example by being committed to their own physical health. They have to infuse physical fitness and healthy living into the organization and lead by example. As the responder's aches and pains add up, and their physical abilities decrease, it begins to chip away at their mental health and identity. Physical health for emergency responders can no longer be optional. The unhealthy responder is the vulnerable responder, and that fact should drive change.

Research on Developing Followers

As discussed in previous chapters, Russell et al. (2015) set forth to conduct a qualitative research study involving the interpretation of company level emergency services officers regarding the role and characteristics of leadership. One of the attributes that emerged to support their theoretical finding of the notion that emergency services leaders must serve the needs of followers was that leaders must develop followers (Russell et al. 2015). Again, the setting of the study took place at large metropolitan fire and emergency services organization in the western United States. Presented below are the results of the study in the words of the study's participants regarding the need that leaders must develop followers.

Develop followers P1 stated that "the front-line officer is the most influential to the crews, they need to educate and always be learning" (P1). P1 went on to argue that functioning within this role "allows the people to do their job without micromanaging them" (P1). P7 added to this by arguing that in order to develop followers a leader must have "a working knowledge of the fire ground and administration" (P7). P2 acknowledged that the officer must possess the "technical knowledge to run a crew, and ensure that tactics are preformed correctly and timely." After acknowledging this, P2 stated that, "the main role of a leader is to be a teacher and mentor" (P2). P2 elaborated on the role of teacher/mentor as "an attitude of learning and improving with personnel management, interpersonal relationships, and communication. It takes a culture of education, continued improvement, and mentoring" (P2). P3 discussed that "leadership is accomplished through mentoring and modeling this service in you so that they can also in turn reflect what they have learned out into the community" (P3). P4 said the leader developing followers into something more "extraordinary and uncommon are the things that inspire others to be more than they are" (P4). P5 stated that a leader "prepares followers to do their job through training followers" (P5). P5 went on to state that a leader is "always teaching his replacement by his actions and coaching" (P5). To add to this, P10 stated that, "a leader within the fire service must have the ability to effectively organize, direct, and mentor followers" (P10).

Summary

As an emergency services leader, you are responsible to grow those who are trusted to your command. These are the individuals that one day will fill vacated leadership positions. A leader's actions and behaviors directly impact each and every one of them. As a leader, one must be laser focused on growing responders; the future of the profession rests with them. The commitment one has to them will be the commitment that they will have to their followers. As with so many other things in the human experience, this too is a cycle.

This commitment to serving the growth of responders needs to be cultivated to a point that it becomes a natural given. The emergency services leader needs to come to terms with the fact that they are first-and-foremost responsible for the mental and physical wellbeing of their followers. Without the responders, there are no emergency services. And unless responders grow into servant leaders, the problems facing the emergency services will remain. Being committed to the growth of people starts with wanting them to be the best that they can be. It means no excuses. As a leader, one must demand and push followers to grow into their full potential.

Case Study A

You have just been appointed the fire chief of a three-station fire department. Something you notice as soon as you assume your position is that your command staff, which consists of a deputy chief of department, two assistant chiefs of operations, and three battalion chiefs, are all of retirement age. In one of your first staff meetings, you bring up the topic of succession. You put these chiefs on the spot by asking them whom they think will be filling their roles once they retire. It becomes clear to you that this department you are now in charge of has never considered succession planning. The command staff informs you that the rank-and-file members lack qualifications for promotion. In fact, two of your three battalion chiefs were hired from other departments. And it gets worse. Your command staff informs you that the rank-and-file of the organization has had discipline issues: a suicide, two line-of-duty deaths in 7 years, a union grievance that has halted the implementation of physical fitness programs, and a city council that decided to cut education and training benefits during the last economic downturn.

Case Study A Questions

1. As the chief of department, how would you begin to address these problems?
2. What do you think are the underlying cause(s)?
3. As the chief of department, what is your primary responsibility?

Case Study B

You have been promoted to the rank of assistant chief of police over human resources and development within a statewide law enforcement agency. Because of past compliance issues, your agency is now using a statewide healthcare company to perform more in-depth physicals of your officers. Previously, the agency relied on an in-house traveling physician to perform general checkups of officers over a year's period throughout the state. In one of your first meetings, you are made aware that at this year's annual physical, more than 35% of your officers statewide were found to have underlying medical problems ranging from pre-diabetes and anxiety issues to heart conditions, sleep apnea, and fatty livers. Furthermore, a majority of your veteran officers are borderline obese and average greater than 25% body fat.

Case Study B Questions

1. How did the agency get to this point?
2. What steps will you take to address this issue?
3. What are the ramifications for addressing and not addressing this problem?

References

Carey, M. G., Al-Zaiti, S. S., Dean, G. E., Sesanna, L., & Finnell, D. S. (2011). Sleep problems, depression, substance use, social bonding, and quality of life in professional firefighters. *Journal of Occupational and Environmental Medicine, 53*, 928–933.

Duckworth, A. L., & Eskreis-Winkler, L. (2013). True grit. *The Observer, 26*(4), 1–3.

Dweck, C. (2006). *Mindset: The new psychology of success.* New York: Random House.

Gallagher, S. (2016). What is enlightenment (and what's in it for me)? *Journal of Consciousness Studies, 23*(1–2), 94–104.

Greenleaf, R. (1977/2002). *Servant-leadership: A journey into the nature of legitimate power and greatness.* Mahwah: Paulist Press.

Groves, K. S. (2007). Integrating leadership development and succession planning best practices. *Journal of Management Development, 26*(3), 239–260.

Hamm, A. O., Weike, A. I., Schupp, H. T., Trieg, T., & Dressel, A. (2016). Affect, empathy, and self-awareness. *Cognitive Brain Research, 17*, 223–227.

Henderson, S. N., Van Hasselt, V. B., LeDuc, T. J., & Couwels, J. (2016). Firefighter suicide: Understanding cultural challenges for mental health professionals. *Professional Psychology: Research and Practice, 47*(3), 224–230.

Hoeger, W. W., & Hoeger, S. A. (2016). *Lifetime physical fitness and wellness: A personalized program.* New York: Cengage Learning.

Hougaard, R., Carter, J., & Coutts, G. (2016). *In one second ahead*. New York: Palgrave Macmillan.

Humphrey, R. (2015). The influence of leader emotional intelligence on employees' job satisfaction: A meta-analysis. In International Leadership Association 17th Annual Global Conference.

Keith, K. (2008). *The case for servant leadership*. Westfield: Greenleaf Center for Servant Leadership.

Kirschman, E. (2004). *I love a firefighter: What every family needs to know*. New York: Guilford Press.

Kirschman, E. (2006). *I love a cop: What every family needs to know*. New York: Guilford Press.

Lopez, S. J., & Louis, M. C. (2009). The principles of strengths-based education. *Journal of College and Character, 10*(4), 1–8.

Marton, F., & Booth, S. A. (1997). *Learning and awareness*. London: Psychology Press.

Mas-Machuca, M., Berbegal-Mirabent, J., & Alegre, I. (2016). Work-life balance and its relationship with organizational pride and job satisfaction. *Journal of Managerial Psychology, 31*(2), 586–602.

Meyer, E. C., Zimering, R., Daly, E., Knight, J., Kamholz, B. W., & Gulliver, S. B. (2012). Predictors of posttraumatic stress disorder and other psychological symptoms in trauma-exposed firefighters. *Psychological Services, 9*(1), 1–15.

Michelson, R. (2006). Preparing future leaders for tomorrow: Succession planning for police leadership. *Police Chief, 73*(6), 16.

Mitani, S., Fujita, M., Nakata, K., & Shirakawa, T. (2006). Impact of post-traumatic stress disorder and job-related stress on burnout: A study of fire service workers. *Journal of Emergency Medicine, 31*(1), 7–11.

Newberg, A., & Waldman, M. R. (2016). *How enlightenment changes your brain: The new science of transformation*. New York: Avery Publishing Group.

Paton, D. (2005). Posttraumatic growth in protective services professional: Individual, cognitive and organizational influences. *Traumatology, 11*, 335–346.

Patterson, K. (2003). Servant leadership: A theoretical model (Doctoral Dissertation). Available from ProQuest Dissertation and Theses Database (UMI No. 3082719).

Rothwell, W. J. (2010). Effective succession planning: Ensuring leadership continuity and building talent from within. AMACOM Division American Management Association.

Russell, E. (2014). *The desire to serve: Servant leadership for fire and emergency services*. Westfield: Robert K. Greenleaf Center for Servant Leadership.

Russell, E., Broomé, R., & Prince, R. (2015). Discovering the servant in fire and emergency services leaders. *Servant Leadership: Theory & Practice, 2*(2), 57–75.

Russo, M., Shteigman, A., & Carmeli, A. (2016). Workplace and family support and work–life balance: Implications for individual psychological availability and energy at work. *The Journal of Positive Psychology, 11*(2), 173–188.

Rutter, M. (1987). Psychosocial resilience and protective mechanisms. *American Journal of Orthopsychiatry, 57*(3), 316–327.

Salovey, P. (1997). *Emotional development and emotional intelligence: Educational implications*. New York: Basic Books.

Salovey, P., & Mayer, J. D. (1990). Emotional intelligence. *Imagination, Cognition and Personality, 9*(3), 185–211.

Satici, S. A. (2016). Psychological vulnerability, resilience, and subjective well-being: The mediating role of hope. *Personality and Individual Differences, 102*, 68–73.

Seigal, T. (2006). *Developing a succession plan for United States Air Forces in Europe fire and emergency services chief officers. Executive fire officer program*. Emmetsburg: National Fire Academy.

Spears, L. (2010). Servant leadership and Robert K. Greenleaf's legacy. In K. Patterson & D. van Dierendonck (Eds.), *Servant leadership: Developments in theory and research* (pp. 11–24). New York: Palgrave Macmillan.

Surjan, A., Kudo, S., & Uitto, J. I. (2016). Risk and vulnerability. In *Sustainable development and disaster risk reduction* (pp. 37–55). Japan: Springer.

Suzuki, W., & Fitzpatrick, B. (2015). *Healthy brain, happy life: A personal program to activate your brain and do everything better*. New York: HarperCollins.

Chapter 10
The Guardian's Cycle of Trust

Eric J. Russell

With contributions by R. Jeffery Maxfield, Jamie L. Russell, and Rodger E. Broomé

> *It is mutual trust, even more than mutual interest that holds human associations together.*
>
> — *Henry Louis Mencken*

Abstract This chapter introduces the concept of the first responder's cycle of trust. This cycle is one that is ongoing and flows between followers, as well as leaders and followers. To operate, emergency service organizations must be built upon trust. Thus, it is imperative for leadership to cultivate a trust-based community. This chapter delineates on the power of trust and how it is held as a core responder virtue.

According to Simpson (2007), trust is simply an individual's willingness to be vulnerable. This idea may sound strange until you reflect upon its meaning. To trust someone means that you are confident that the person will be credible, authentic, and able. This makes you vulnerable in the relationship if that individual turns out to be less than trustworthy. Trusting someone means that you can be betrayed or wronged by that person. Thus, when you trust, you willingly make yourself vulnerable to another.

On the other hand, the individual receiving the gift of trust is also vulnerable. They are vulnerable because you trust them. They are burdened with the pressure that comes from being trusted. This has to do with the fact that the trusted individual is required to always do the right thing even when it is difficult. It needs to be understood that the trusted individual squanders the gift of trust when they fail to be true.

This concept of vulnerability of the trusted agent has nothing to do with failure as long as it is free of malice. The expectation that the trusted individual can never fail is unrealistic. The only expectation is that the failure is unintentional or accidental. One cannot be called a servant leader if they place the burden of perfection on the shoulders of others.

Again, as it has been discussed in other areas of this book, responders are vulnerable because they are human. Understanding this truth leads to developing trust-based relationships that strengthen the responder's wellbeing. It begins by realizing

E. J. Russell, *In Command of Guardians: Executive Servant Leadership for the Community of Responders*, https://doi.org/10.1007/978-3-030-12493-9_10

that vulnerability is not weakness; the vulnerable responder is the healthy responder (Simpson 2007). Because to be the vulnerable responder means that they are present in their true self, free of the masks that so many wear to protect them from harm (Brown 2015; Rousseau et al. 1998).

What is understood about trust is its ability to foster risk taking (Colquitt et al. 2007). What this means for the emergency services is the more trust that exists, the more risk responders are willing to take. On the surface, one may think about bravery or risking one's life, and such reflection is not wrong. However, this concept of trust in risk taking goes beyond the obvious emergency scene. The profession needs to operate in trust; the job is inherently about being in harm's way which is a given. Where risk is needed is beyond the emergency. For example, bringing forward a new idea, presenting a process, or being open and honest about what one is going through. These are only a few examples of risk beyond the emergency scene and they can only occur when individuals trust others to be open and vulnerable (Simpson 2007).

The goal of this chapter is to unpack this concept of trust within the emergency services because the profession needs to stand on a foundation of trust. It is impossible to carry out the functions of emergency response without reciprocal trust-based relationships, and in this case the professions embody trust when it comes to operations. However, trust needs to flow in all directions as an ongoing cycle, and it needs to exist in the everyday mundane. Thus, the trust that seemingly comes so naturally during chaos needs to flow over to non-emergency situations.

Trust of Self

The cycle of trust begins with each individual having a trust of self. This involves the trust for one's own capabilities and character (Lehrer 1997). Trust has to begin with trusting self because if you cannot trust yourself you cannot trust someone else. If one reflects upon the concept of vulnerability as it pertains to trust, the individual that does not trust in self cannot receive trust from another. For if one does not trust in self, receiving trust from another creates an existential crisis. This has to do with the individual not being able to accept the trust of another because they don't trust that they themselves deserve to be trusted (Lehrer 1997).

Trusting in self has nothing to do with arrogance. Rather, personal trust is about believing in self. The ability to trust in one's self begins with believing that one is credible and capable (Lehrer 1997). Again, not by being arrogant or cocky, but rather, confident in one's own abilities and character. On the surface the concept of trusting in self seems to be a given. However, it is not a given, but rather, a cultivated virtue that comes from the responder honing their skills and proving to themselves that they are capable of doing the work.

The responder that is unsure of their own abilities and limitations subconsciously knows that they are a danger to self and others. Thus the responder that cannot trust in self cannot receive trust nor trust others. Leaders in the emergency services have to adhere to training standards and assessments of individuals so that those individuals can learn to trust in their own abilities.

Trust Between Responders

As noted earlier in the text, the term servant leadership can at times confuse people. The notion of servant leadership is far more than just a leader over people. Servant leadership is about being. Therefore, the characteristics, constructs, and attributes of the philosophy are more than being over others. This is why it is important to discuss the characteristics of trust as they relate to the follower relationship within the emergency services. Because in order for responders to successfully function with one another, they have to be in service to one another and above all, they have to be able to trust in each other's abilities.

The emergency services professions operate on a team structure (Jouanne et al. 2017). Every function from ventilation and interior attack scenarios on the fire ground, to law enforcement responses to active shooters and suspicious vehicle traffic stops all play out as a team. The concept of the lone-responder is one of myth, a fictional character in a movie. The essence of emergency response lives in the trust-based relationships between followers.

As it is with any other relationship, trust between followers is not automatic. This trust-based relationship is something that is cultivated over time (Simpson 2007). It does not exist because one person says "you can trust me" and the other agrees. Instead, the trust-based relationship between followers, one that is vital for operating under emergency conditions, is forged out of each responder's being. This means that the trust-based relationships between followers only comes to fruition after each responder proves that they are trustworthy, thus earning the trust of others (Politis and Politis 2017).

This trust-based relationship is something that is constructed over time (van der Werff and Buckley 2017). It comes from proving one's ability so that they can be trusted. It originates from operating with the other responders on emergency scenes and in chaotic situations. This trust-based relationship between followers occurs because individuals have proved themselves. This is why the emergency services career field uses probationary periods and rookie positions, so that newly appointed responders that have a desire to serve can prove themselves worthy of trust. This time period in the responder's career is where they earn the respect of their fellow responders. During one's probationary period, they are under a microscope because they are asking their fellow responders to gift to them trust and willingness to rely on them in times of danger.

Once the individual responders earn the trust of their peers, their peers welcome them into the family unit. It is at this point that responders can let down their armor and be human with their peers. It is that trust that one can only earn through operations and experiences that allow for personal friendships and bonds to take hold. And it is these friendships and bonds between responders that allow them to safely operate and flow as a team. Furthermore, it is these relationships that allow responders to be open and connected with others, thus strengthening resiliency and post-traumatic growth (Anderson 2018; Paton 2005). After all, there is something to be said about trust-built relationships forged from being in harm's way with another.

Those who find themselves in command positions have a responsibility to cultivate these relationships (Moorman et al. 2018). With that said, it needs to be understood that they cannot force nor demand trust. However, what they can do is build assigned teams that stay together for a while, because, as noted earlier, these relationships only form over time and through multiple collective experiences. Followers need to have the opportunity to work together. In addition, leaders need to be willing to put forth team-building opportunities in training and in-service scenarios. Again, a leader can't demand that followers trust one another—that only occurs on the interpersonal level. Yet they can do everything in their power to help foster and strengthen these critical relationships. It begins by keeping crews together so that they can form a bond, which leads to a willingness to be vulnerable with each other (Davis 2017; Rousseau et al. 1998; Simpson 2007).

The Leader-Follower Relationship

If one were to ask the average person to reflect upon the concept of trust within organizations, a majority would reflect upon the leader-follower relationship (Hurley 2006). Specifically, most would reflect upon the mistrust they have for their leaders (Hurley 2006). This is a foundational problem effecting the health and wellbeing of organizations. If your people are barely functioning, or worse yet, surviving, in an environment void of trust, eventually the system begins to fail (Engelbrecht et al. 2017). When there is a lack of trust between those leading and those being led, the organization begins to rot due to limited communication and openness.

Acknowledging this issue naturally leads one to ask, why does this lack of trust exist within organization? One possible answer is that the behavior of leaders produces an atmosphere of mistrust (Engelbrecht et al. 2017; Jaiswal and Dhar 2017). It needs acknowledging that this behavior often times occurs out of self-preservation, not malice. Many individuals serving in a leadership capacity are in a state of self-preservation, and again, this is due to organizational environments and culture. Often, those in leadership positions find themselves navigating problems and push-back from above. This, in turn, places the leader in a situation where they are required to serve the bidding of their superiors. Because of pressures to serve those above them, leaders begin to see their followers as problems or liabilities (Bligh 2017). Leaders then fake their credibility and commitment when followers are present, yet at the same time, they keep them at a psychological-arm's length so as to avoid letting them in and serving their needs (Engelbrecht et al. 2017; Mo and Shi 2017).

A servant leadership approach offers a possible solution to overcoming this toxic issue. It does so by forging relationships of open dialogue and trust based upon the integrity of the leader (Maxfield and Russell 2017). In a servant leader-follower trust cycle, the leader has built a community of trust and openness with their followers. They are honest about what is coming down from the top and why. When they speak within this community they are truthful. This behavior strengthens a leader's credibility and creates the cornerstone of trust. This begins with being straightforward about one's role in a specific issue.

A leader needs to openly acknowledge where their authority ends. At times, emergency responders will desire their leader to carry the torch. They conceive the officer as the one with the power to make change or pushback against those above. The officer in this case must make it clear that they are also a follower, and at times, powerless in the face of real authority, for even the chief of department answers to someone above them. When the leader is honest about their limitations it gives them credibility with followers. The majority of followers can then see their leader as someone who has to navigate the same situation. They then desire to support and serve the leader as a group instead of developing into an "us versus them".

Again, this can only happen with open dialogue and credible behavior. Outside powers and influence will attempt to "other" the leader-follower relations and undermine the trust. The only power that can hold this back is honesty and transparency.

Understanding the trust-based relationship is vital to emergency services delivery. In the chaos of the incident, followers need to trust in their leader's ability to make decisions and give orders. In turn, leaders need to be able to trust that followers that receive the orders will carry them out. The entire emergency operation relies on this relationship (Russell 2015). When followers develop mistrust for their leaders, they hesitate, or worse, they simply ignore orders and do things their way. Such behavior is not obvious. Instead, it comes in the form of not hearing certain radio traffic, or hearing something that was not said. Such behavior is a symptom of leaders abusing the trust of followers (Schweitzer et al. 2006).

Mistrust is also an issue beyond the emergency scene. For many agencies it is even a bigger issue because of the para-military structure. When an emergency occurs, many responders are able to look past the personal issues and carry out the work in a safe and orderly manner. Even when relationships between leaders and followers are on shaky ground, most can come together for the good of both the team and the public. This, however, is not the case when it comes to the non-emergency scene. This is where toxic leader-follower interactions can have a detrimental impact to the organization (Schweitzer et al. 2006). The noxious relationship and the atmosphere of mistrust make it improbable that a leader's vision will be carried out. Followers navigate the system doing the minimum required, function only within the parameters of their job description, and become unwilling to go above-and-beyond (Mo and Shi 2017; Politis and Politis 2017).

The leader that is willing to recognize and repair these relationships understands the power they will be gifted from followers. Most people, including first responders, desire to work for honest people. Working in a state of mistrust is detrimental to ones being. Moreover, most leaders want to get along with their people and they want to be able to trust and to be trusted. It simply makes work enjoyable. The servant leader is aware that legitimate power is found in trust.

The legitimate power gifted to the leader comes in the form of an ease of persuasion and followers desiring to be empowered. As Russell et al. (2017) found it is in the self-interest of the leader to have followers willing to trust them. Trusted leaders find that they do not have to spend time on the mundane, or selling an idea. Because followers trust their leaders, they desire to carry out the vision and they wish to be delegated to. To put it another way, they want to help. A trusted leader benefits simply by being trusted—their burdens are lessened by those whom they have earned trust (Russell et al. 2017).

Trust as a Core Value

Maxfield and Russell (2017) set forth to discover how leaders within the emergency services interpreted the lived experience of becoming a leader. Emerging from their study were themes associated with personally held values of integrity, honesty, and trust (Maxfield and Russell 2017). Each participant in the study discussed similar sets of values that guide him or her as leaders. Each participant highlighted different aspects that make up the construct of trust, such as honesty and integrity (Maxfield and Russell 2017).

This finding goes to what Russell (2015) argued to be the factors that bring responders to the career field, establishing their desire to serve. Moreover, it is the commonality between the emergency services and the philosophy of servant leadership that comes to light in leadership research within the professions (Maxfield and Russell 2017; Reed 2015; Russell et al. 2015; Russell 2015; Sedlmyer 2017). Because of empirical works on the subject, we know that trust is the cornerstone not only during emergency operations, but also, within the relationships that exist in the profession.

Fostering the Guardian's Cycle of Trust

Each aspect discussed in this chapter comes together to form what is known as the cycle of trust. It is imperative for this cycle to exist in order to carry out safe and effective emergency operations. The cycle of trust begins with individuals' willingness to be vulnerable, to open themselves up to both self and others, and to be aware that such vulnerability can cause personal harm.

The responder's cycle of trust continuously flows through three states: self-trust, trust between followers, and trust between leaders and followers. The first is self-trust, and that is simply the responder believing in their abilities and character. The emergency services leader is responsible for cultivating self-trust within followers. This is accomplished through one's commitment to the growth of one's people. Here the servant leader nurtures self-trust through training and educational opportunities; putting followers into controlled chaos and scenarios that allow them to believe in their abilities. As discussed earlier, when an individual lacks self-trust it creates an existential crisis because they themselves cannot receive trust. However, when the individual has self-trust then they are psychologically willing to be trusted. The emergency services leader needs to be aware that when followers lack self-trust, they cannot trust others nor can they be trusted. This is a dangerous situation because it puts operations and the responders at risk.

Second, once these individuals who possess self-trust come together, then the trust-based relationships between followers can form. This aspect of trust is the cornerstone of emergency operations. The trust-based relationships between individuals working under time-pressure and consequence are essential. In the moment of chaos, individuals cannot question whether they can trust an individual working on the scene with them.

For the emergency services leaders, fostering this area of trust begins by building healthy communities for responders. The positive community allows responders to get to know each other personally, leading to trust-based esprit-de-corp. This community is where they live, train, and respond together. This environment allows for a natural understanding of each responder's abilities and weaknesses. They organically get to know and trust the members of their team. Cultivating this community ensures that when the emergency occurs, responders seamlessly work together because of a genuine trust in the abilities and character of others.

The third aspect of this cycle is the trust between leaders and followers. It is in this part of the cycle where trust flows in both directions, followers trusting in their leaders and leaders trusting in their followers. Within many organizations, this is an issue, one that holds back great achievement and success. Though the trust between leaders and followers flows both ways, the responsibility of cultivating this relationship falls predominately on the shoulders of leaders due to positional power. Based solely on position they have a say in their future and their income. Therefore, it is up to leadership to go beyond the given positional aspects of their role to earn the trust of followers.

This begins with being an authentic and credible leader—one who is open and honest. Followers learn to trust the leader and because of that trust, gifts to them legitimate power. It is important to understand this because one's positional power is not legitimate power; it is simply a place on an organizational chart. To be a real leader, followers must desire to follow you, and when they desire to follow you, it means they trust you.

The states of self-trust, trust between followers, and trust between leaders and followers make up the responder's ongoing cycle of trust. The professions thrive within this cycle because it is impossible for healthy relationships to exist without trust. It is the responsibility of those in command of guardians to ensure that it continues to flow for the sake of both responders and the organizations in which they serve.

Summary

The purpose of this chapter was to introduce and delineate on the different aspects that make up the guardian's cycle of trust. The concept of trust is the cornerstone of emergency service professions. On the emergency scene, trust is what allows for effective operations. In the day-to-day activities and living conditions among responders, trust is what forms bonds and healthy relationships. Trust is what allows responders to believe in and seek out the council of their leaders.

The cycle of trust that flows between followers, and between leaders and followers, is an ongoing healthy state of being. For those who find themselves in command of guardians, it is imperative for them to cultivate a trust-based community of responders. Trust is what brings about a leader's legitimate power, a power that only comes to a leader as a gift from followers.

Case Study A

You have been appointed Chief of Fire and Emergency Services over a large federal fire department. The department consists of 120 members and four geographically located stations. During you first week, two of your stations respond to a structure fire in base housing. You decided to self-dispatch to the incident without announcing it over the air just to observe the operation leaving your assistant chief of operations in command. As you respond, you listen to radio traffic of companies working on scene. The operation seems to be running smooth and within minutes the interior fire attack captain reports a loss-stop and requests to move to an overhaul operation. The assistant chief in command of the scene orders the safety officer to enter the structure to confirm what the fire attack captain reported. In addition, you hear the incident commander micromanaging the ventilation operations. When you arrive on scene, the incident commander sees your vehicle and asks over the radio if you want to be passed command. You reply that you are just observing and that he should remain in command. Later that day, the assistant chief that was in command of the incident comes to your office and asks to speak with you. You agree and as he enters he closes the door. He says that he did not appreciate being spied on by his chief and that he feels as if you did not trust him to do his job.

Case Study A Questions

1. What problems do you see with this assistant chief?
2. How does this assistant chief impact your organization?
3. What steps can you take to help this assistant chief?

Case Study B

As a Lieutenant Colonel appointed to command an investigation unit within a military law enforcement operation, certain issues have come to light. In the last year, three of your investigators have been accused of doctoring evidence and are now under investigation by the Inspector General's office. Investigators from the Inspector General's office have made you aware they have found damning evidence against your officers. All three of these incidents are separate and involve unrelated cases. The Judge Advocate General's office uses the work of your office to charge individuals and groups with crimes. Until today, this investigation was sealed and the public had yet to be made aware. However, this morning you received a call from a national news outlet looking for comment from your office regarding this situation. The reporter stated that they have documents linked to the investigation and that the story will be published within the next 24-h.

Case Study B Questions

1. How does this incident impact trust within your agency?
2. How does this incident impact trust between your agency and those you work with?
3. What are the first steps you would take after hanging up with the reporter?

References

Andersson, T. (2018). Followership: An important social resource for organizational resilience. In *The Resilience Framework*. New York: Springer.

Bligh, M. C. (2017). Leadership and trust. In J. Marques & S. Dhiman (Eds.),. (2017) *Leadership today: Practices for personal and professional performances, VI* (pp. 21–42). New York: Springer.

Brown, B. (2015). *Daring greatly: How the courage to be vulnerable transforms the way we live, love, parent, and lead*. New York: Penguin.

Colquitt, J. A., Scott, B. A., & LePine, J. A. (2007). Trust, trustworthiness, and trust propensity: A meta-analytic test of their unique relationships with risk taking and job performance. *Journal of Applied Psychology, 92*(4), 909–927.

Davis, N. (2017). Review of followership theory and servant leadership theory: Understanding how servant leadership informs followership. In *Servant Leadership and Followership* (pp. 207–223). New York: Palgrave Macmillan.

Engelbrecht, A. S., Heine, G., & Mahembe, B. (2017). Integrity, ethical leadership, trust and work engagement. *Leadership and Organization Development Journal, 38*(3), 368–379.

Hurley, R. (2006). The decision to trust. *Harvard Business Review, 9*, 55–71.

Jaiswal, N. K., & Dhar, R. L. (2017). The influence of servant leadership, trust in leader and thriving on employee creativity. *Leadership and Organization Development Journal, 38*(1), 2–21.

Jouanne, E., Charron, C., Chauvin, C., & Morel, G. (2017). Correlates of team effectiveness: An exploratory study of firefighter's operations during emergency situations. *Applied Ergonomics, 61*, 69–77.

Lehrer, K. (1997). *Self-trust: A study of reason, knowledge, and autonomy*. Oxford: Clarendon Press.

Maxfield, J., & Russell, E. (2017). Emergency services leadership: The lived experience. *The Journal of Homeland Security Education, 6*, 1–13.

Mo, S., & Shi, J. (2017). Linking ethical leadership to employee burnout, workplace deviance and performance: Testing the mediating roles of trust in leader and surface acting. *Journal of Business Ethics, 144*(2), 293–303.

Moorman, R. H., Blakely, G. L., & Darnold, T. C. (2018). Understanding how perceived leader integrity affects follower trust: Lessons from the use of multidimensional measures of integrity and trust. *Journal of Leadership & Organizational Studies*, 2018.

Paton, D. (2005). Posttraumatic growth in protective services professional: Individual, cognitive and organizational influences. *Traumatology, 11*, 335–346.

Politis, J. D., & Politis, D. J. (2017. December). The role of servant leadership on interpersonal trust and performance: The mediating influence of interpersonal trust. In 13th European Conference on Management, Leadership and Governance: ECMLG 2017 (p. 382).

Reed, L. (2015). Servant leadership, followership, and organizational citizenship behaviors in 9-1-1 emergency communications centers: Implications of a national study. *Servant Leadership: Theory and Practice, 2*(1), 71–94.

Rousseau, D., Sitkin, S., Burt, R., & Camerer. (1998). Not so different after all: A cross-discipline view of trust. *Academy of Management Review, 7*, 393–404.

Russell, E. (2015). *The desire to serve: Servant leadership for fire and emergency services.* Westfield: Robert K. Greenleaf Center for Servant Leadership.

Russell, E., Broomé, R., & Prince, R. (2015). Discovering the servant in fire and emergency services leaders. *Servant Leadership: Theory & Practice, 2*(2), 57–75.

Russell, E., Maxfield, J., & Russell, J. (2017). Discovering the self-interest of servant leadership. *Servant Leadership: Theory & Practice, 4*(1), 75–97.

Schweitzer, M., Hershey, C., & Bradlow, E. (2006). Promises and lies: Restoring violated trust. *Organizational Behavior and Human Decision Processes, 101*(1), 1–19.

Sedlmeyer, L. R. (2017). Fire officer leadership strategies for cost management (Doctoral dissertation, Walden University).

Simpson, J. (2007). Psychological foundations of trust. *Current Directions in Psychological Science, 16*(5), 264–268.

van der Werff, L., & Buckley, F. (2017). Getting to know you: A longitudinal examination of trust cues and trust development during socialization. *Journal of Management, 43*(3), 742–770.

Chapter 11
Fostering Responder Servant-Followership

Eric J. Russell

With contribution by Jamie L. Russell

> *Followers are more important to leaders than leaders are to followers.*
>
> —*Barbara Kellerman*

Abstract This chapter reflects upon the concept of servant-followership within the community of responders. Throughout the emergency services, the lowest paid and least-esteemed positions are also the most dangerous. These are the quintessential followership positions consisting of the firefighter on the nozzle, the patient technician, and the patrol officer. Each of these positions is where the rubber meets the road, most loss is realized, and the most injury and line of duty deaths occur. These follower positions make up the bulk of the emergency services career fields. The vast majority of the men and women who serve within the emergency services find themselves at this level. As noble as these positions are, textbooks and writings seemingly disregard their importance by concentrating on the role of the leader. It is the aim of this chapter to bring to life the role that followership plays in the emergency services and how executive leaders can foster healthy followership within their organizations.

One of the three pragmatic questions forming Greenleaf's (1977/2002) philosophy of servant leadership goes to the heart of this chapter. The question asked whether followers "while being served, become healthier, wiser, freer, more autonomous, more likely themselves to become servants" (Greenleaf, 1977/2002)? This question is one that emergency service leaders need to continuously reflect upon because of two distinct leadership challenges facing leaders when it comes to followers. The first being, are followers being led in a way that allows them to reach their full potential? The second asks, are followers being led in a way that will allow them to one day become servant leaders?

The purpose of this chapter is to delineate on the constructs of servant followership identified by Winston (2003), presenting the relationship each has to the emergency responder in order to address those two distinct challenges. According to

E. J. Russell, *In Command of Guardians: Executive Servant Leadership for the Community of Responders*, https://doi.org/10.1007/978-3-030-12493-9_11

Winston (2003), the first five constructs of the servant-follower are *agapao* (love), commitment to the leader, self-efficacy, intrinsic motivation, and altruism towards the leader/leader's interest (Winston 2003). The sixth construct, service, becomes the summary for this chapter to show how the constructs come together. The goal of addressing servant-followership is to give those in command of guardians an idea of what specific constructs they need to focus on and cultivate within responders. The aim is so that responders can grow in their abilities to serve, as well as become servant leaders one day.

Before this chapter moves into the specific constructs of servant followership, it needs noting why this understanding is important. Currently, there are 4,544 uniformed and sworn career and mostly career fire departments in the United States employing 345,600 career firefighters (Hylton and Stein 2017). These career and mostly career fire departments comprise just 15% of all fire departments operating in the United States (Hylton and Stein, 2017). For law enforcement, if you included federal, state, and local agencies, there are 18,000 police departments in the United States employing 929,000 uniformed and sworn officers (Reaves 2011). This means that in the fire service there are 4,544 fire chiefs, making up less than 1.4% of all career firefighting positions. For law enforcement, this means there are 18,000 police chiefs, making up less than 2% of all law enforcement positions. Therefore, roughly 98% of all career personnel function in some type of followership position within their organization (Griffin 2017).

The word leadership is one of the most overused, misunderstood words in our vernacular. Individuals complain about leadership. When something goes wrong they demand changes in leadership. Those in power or those desiring power argue for the "great man" to be a leader, doing their best to convince the masses that a single person can change everything. It seems as if the modern understanding of leadership is based upon myth. It is possible that this myth is the reason so many people believe leadership is more than it really is, and it may have something to do with the entertainment and literary worlds. In the world of fiction, books and movies rarely hold up followers.

In addition to the fictional entertainment and literary genres, leadership and management books for the emergency services also focus on leadership and seemingly ignore followership, passing over its vital importance within the career field. Moreover, the texts that support the development of the responder focus solely on ones doing and ignore ones being. This is why the constructs of servant-followership needs to be discussed in leadership books, because without followers, there isn't a need for leaders (Martin 2008).

Agapao (Love)

We understand from past writings that from a servant leader perspective, leaders love their people (Greenleaf 1977/2002; Winston 2003). From that moral love, the servant leader sets forth to serve the needs of their people so that, in turn, those individuals can serve the leader and the organization. This idea of moral love from the follower standpoint involves the follower developing a moral love for his or her leader. From that love, the follower then willingly carries out the leader's orders and vision.

It is from a moral love that the relationship of the leader-follower grows because when something grows out of love it is virtuous. From 35,000 feet you can see where this construct strengthens the notion of trust between people. For the emergency services, understanding this concept is vital for strong followership to take hold within the career field (Soler 2017).

For example, the emergency services leader needs to be able to set a vision for the organization, as well as give orders to be carried out on the emergency scene (Álvarez et al. 2014; Soler 2017). Great followers, who have developed a moral love for their leaders, have little issue carrying out said orders (Moore 2008). Because they developed a moral love for their leaders, they in turn desire to serve the needs of their leaders, going beyond obligation because of position. This idea encompasses both the emergency and the non-emergency situation.

A follower's moral love for their leader grows as their leader's moral love for them grows. This is what Winston (2003) was identifying as the circular role of moral love in the leader-follower relationship. Because this is a circular ideal, successful follow-ership and successful leadership both have to begin with a moral love for one another. The leader-follower relationship within the emergency services has to harness this moral love on both sides. Once that is done, greatness can come from it. When there is a moral love on both sides, there is an understanding that the needs of each other matter, and both parties willingly desire to serve said needs (Buchanan 2007).

Those in leadership positions have an obligation to morally love their followers. However, they do not have a right to be loved. Like trust, the love of followers is earned. Therefore, it is the responsibility of the leader to cultivate the relationship. It begins by being an ethical leader. It comes from being just and caring. Love cannot be demanded from those who follow. For the virtuous leader, this path is not a difficult one and it leads to follower commitment.

Commitment to the Leader

Cowboy poet Paul Harwitz wrote a poem title "Ride for the Brand" The premise of the poem has to do with the loyalty of the follower to the brand that they are serving. In his poem, a young man asks a more experienced cowboy about what riding for the brand means. The older cowboy takes the time to explain what it means to him, things such as working hard, having loyalty, being trustworthy, having respect for others, taking only what is needed, and giving the best of self. Looking at this poetic concept through an emergency services lens, it is a loyalty to the brand in which the individual is working for. The understanding of the brand is the branding that is put on the cattle and the horses owned by a certain rancher. The follower, in this case a cowboy, rides for that specific brand; they are loyal to that rancher.

If we use this premise as a teaching moment for being committed to one's leader, it is commitment to that individual or individuals that you serve as a follower. It means that you, as a follower, understand that your role is carrying out their orders and bringing their vision that you have been persuaded to believe in to fruition (Bennett 2017). This is the idea behind the commitment to the leader. Within the servant leader-follower relationship, we understand that the leader has to be committed to the follower; he or she needs to serve their needs so that the follower can grow. Yet in return, the follower needs to be committed to the needs of the leader. In this case, it is doing the things that the leader needs you to do (Havins 2011). Moreover, it is acting in a way that does not bring shame upon the organization or your leadership. A follower's commitment to the leader is one that protects the leader by being committed to their leadership and respecting their position and authority (Spell 2009).

For the emergency services, being committed to one's leader means being committed to their orders and their authority (Tobia 2011). A commitment to the leader is a commitment to the brand, meaning, a commitment to the position of authority they hold in the organization where you both serve. Within the emergency services, a commitment to a leader is more than just a commitment to an individual. It is a commitment to the organization's company and chief officers. Like love, this commitment is earned by being committed to followers. It is then reciprocated back to the leader.

Self-Efficacy

The notion of self-efficacy is the idea that a follower has honed his or her skills and abilities to a point that they believe they can accomplish the task. Winston (2003) noted, "Self-efficacy is the follower's perception of what the follower can and cannot do in terms of his/her capability" (p. 5). In order to be a servant-follower, this construct is vital to understand. At its core is a follower's commitment to be the best that they possibly can be so that they can serve their leader and their organization's needs. It means that followers desire to grow in their knowledge, set-aside time to learn, and go above and beyond the basic requirements of their positions (Soler 2017).

For the emergency services, the self-efficacy of the emergency services responder is vital to meeting the needs of the organization's mission. When followers do not grow in their abilities, when they fail to see that the better they are the better the organization is, and therefore, the better the operation is, they miss out on an opportunity. For instance, self-efficacy would drive a follower to want to pursue advanced certifications above and beyond the basic job requirements (Griffin 2017). It means that they would want to pursue both an education and training so that they can fill the positions above them as needed. The consciously aware follower wants to be an individual that leaders know they can rely on, empower, and delegate to (Bennett 2017).

The emergency services are responsible for so many areas. Followers that understand self-efficacy will serve their leaders as well as their organization by becoming an individual that can handle multiple responsibilities. Servant followers are members of the rank-and-file that desire to serve the needs of their leaders by becoming an individual that their leaders can turn to (Gilbert and Whiteside 1988; Tobia 2011). The servant follower is a person that understands that as their leaders serve their needs they in turn practice self-efficacy so they can serve their leader's needs. This strengthens the leader-follower relationship from both sides, each acting in service to the needs of the other (Gilbert and Whiteside 1988).

Intrinsic Motivation

The concept of intrinsic motivation is the individual's desires to be better, to achieve goals, to do something extraordinary. The idea behind intrinsic motivation is that the servant-follower is internally motivated to be a better self. As Winston (2003) noted, "This inward propensity is not responsive to external rewards or threats but internally focused on the individual's desires" (p. 6). In this construct, the servant-follower is one that is motivated by the desire to serve both their leader and organization and in doing so they succeed. What Winston (2003) is saying is that this motivation is an internal drive coming from a follower's inner desire to be the best at what they do and achieve.

The concept of intrinsic motivation is one that greatly benefits both the leader and the organization because the servant-followers within that organization are motivated to carry out the vision (Martin 2008). Part of intrinsic motivation is finding pleasure in one's role. As Winston (2003) wrote, "Intrinsic motivation results in pleasure for the follower in doing the task" (p. 6). We see this concept in so many self-help books and leadership texts. This idea that the individual can make meaning and find happiness in what they do professionally. This is the key difference between a job and a career. For a career, it is defining for the individual. As discussed in an earlier chapter, this concept especially holds true for the emergency services responder; the profession defines the individual's identity.

Professional responders find meaning in their work; therefore, being an emergency services professional means more than simply doing a job. The notion of being a professional responder outwardly expresses who one is as a person. Being a professional and having a career in the emergency services, means that the individual is motivated to be better tomorrow than they were today (Álvarez et al. 2014). They personally desire to grow physically and mentally stronger. They take on projects that can better the organization, projects that all so often do not come with monetary reward, but rather, inner peace and accomplishment.

The intrinsically motivated servant-follower is an individual that makes their leadership, as well as their organization, better. They find meaning in their work and from that meaning they are able to serve their leaders. The intrinsically motivated follower finds pleasure and makes meaning from their efforts

Altruism Towards the Leader

The concept of altruism within the leader-follower relationship has to do with how the individuals in their roles interact with one another. The altruistic leader gives of himself or herself to the follower within the leader-follower relationship. This is done so that the follower can have their needs met in order to grow and self-actualize. In order for this to not be a one-sided relationship, Winston (2003) noted that the servant-follower must also be altruistic with their self to their leader in order for that cycle to work. This creates a continuous cycle of giving from leader to follower and from follower to leader. It is vital to understand this concept as well as understand that this altruism doesn't exist outside of that relationship; however, it is fundamental within said relationship (Russell 2016).

Within the emergency services, leaders give of themselves to their followers so that their followers can actually do the work. Leadership spends time educating, mentoring, and finding opportunities for followers to grow. In addition, they make the case for their followers' needs to those above. In turn, followers within the emergency services profession give of themselves to their leaders.

Summary

That cycle of altruistic giving between the leader and follower flows into a desire to serve one another. As it is with the leader serving the needs of the follower, the servant-follower desires to be in service to the leader. This desire to serve the leader comes from a love they have developed for their leader. Because of this love, they become committed to their leader, they practice self-efficacy, and become intrinsically motivated. These constructs flow between the follower and the leader (Winston 2003). The emergency services benefits by this relationship. As followers are motivated and committed, organizational leadership can take comfort knowing that those individuals have the best interest of the organization in mind (Martin 2008). Additionally, they understand that the goals of the organization and the vision that has been set forth will be carried out. It is this cyclical relationship of the leader and servant follower that is unlike anything else due to its ability to strengthen an organization so that everybody benefits. And it all begins from a common desire to serve.

Case Study A

You are the fire chief of a large international airport fire department. Because of changes in airport leadership, your organization is under the microscope and it is possible that that services will be contracted out. The new airport executive comes from another international airport where fire and emergency service operations were carried out by a local municipal jurisdiction. This executive sees cost savings from outsourcing emergency services and views fire and rescue operations as a monetary black hole. Airlines pay landing fees, some of which go to fund fire and rescue operations; however, at best, these funds break even at the end of each fiscal year. Within the fire and emergency services, the only area that can generate consistent revenue is with billing insurance for emergency medical services.

Currently, your agency operates at a basic life support level and to make matters worse, advanced life support comes from a local jurisdiction located off the airport. Your department averages 2,600 emergency medical service calls annually with 56% of them needing advanced life support and 70% requiring hospital transport. You propose, to the new executive, that your organization become an advanced life support agency, staffing two paramedic ambulances. In the proposal, you ask for funding for a billing system, an allowance to pay for 14 members of the organization to attend a university paramedic program, and an allocated overtime fund to pay for their absences. You show on paper based upon historical call volumes and run reports that the agency can make money after 17 months. The airport execute agrees with the caveat that if the EMS program does not generate a profit to the airport after 3 years, he will pursue outsourcing.

Case Study A Questions

1. How can servant followers make this vision come to life?
2. What role does a follower's intrinsic motivation play?
3. How would you go about persuading your followers to see your vision?

Case Study B

You have just been sworn in as the assistant chief of training and education for a port authority police department. On your first day, you were informed that you were hired because not a single member of the department possessed the qualifications to fill the position that you now hold. When you asked why, you found out that union executives demanded members not pursue qualifications until they were promoted into and paid at the position that requires it, or that the port authority pays each member more for each qualification they hold above and beyond what is required for their current position.

That week, you call a meeting with a union representative to discuss this issue. The union president explains to you that past leaders have used members in positions and not paid or promoted them as a way to save the port authority money. He gives you an example from several years ago of a police corporal being made to fill in as a lieutenant for over a year and never compensated or credited for the additional work and responsibility. He tells you that because of port authority policies, members must hold the qualifications needed to perform work at a certain level. He then informs you that the only way to protect the members from being used in higher positions was to enact this union bylaw regarding not perusing higher qualifications.

Case Study B Questions

1. What is the breakdown between leaders and followers?
2. Can you improve this situation?
3. What role can developing servant-followers play in preventing this situation from reoccurring?

References

Álvarez, O., Lila, M., Tomás, I., & Castillo, I. (2014). Transformational leadership in the local police in Spain: A leader-follower distance approach. *Spanish Journal of Psychology, 171, E42.*

Bennett, C. W. (2017). Followership. *Police Chief, 62*(9), 28–32.

Buchanan, E. (2007). Can there be leadership without followership? *Fire Engineering, 160*(8), 105.

Gilbert, G. R., & Whiteside, C. W. (1988). Performance appraisal and followership: An analysis of the officer on the boss/subordinate team. *Journal of Police Science & Administration, 16*(1), 39–43.

Greenleaf, R. (1977/2002). *Servant-leadership: A journey into the nature of legitimate power and greatness.* Mahwah: Paulist Press.

Griffin, D. (2017). We're all followers before leaders. *Firehouse, 42*(7), 70–71.

Havins, M. J. (2011). An examination of the relationship of organizational levels and followership behaviors in law enforcement. *Dissertation Abstracts International Section A,* 71, 4081.

Hylton, J., & Stein, G. (2017). *U.S. fire department profiles* [online]. Accessed from: http://www.nfpa.org/-/media/Files/News-and-Research/Fire-statistics/Fire-service/osfdprofile.pdf

Martin, R. (2008). Followership: The natural complement to leadership. *FBI Law Enforcement Bulletin, 77*(7), 8–11.

Moore, F. (2008). Follow the leader. *Wildfire, 17*(4), 16–18.

Reaves, B. (2011). *Census of state and local law enforcement agencies (U.S. Department of Justice, Office of Justice Programs. Bureau of Justice Statistics, NCJ 233982).* Washington, DC: U.S. Government Printing Office.

Russell, E. (2016). Servant leadership's cycle of benefit. *Servant Leadership: Theory & Practice, 3*(1), 52–68.

Soler, L. (2017). Followership: An essential component of leadership. *The Police Chief,* (1), 24.

Spell, J. (2009). Follow the leader, redefined. *Fire Chief, 53*(12), 26–29.

Tobia, M. (2011). The back step. The paradox of followership. *Fire Rescue Magazine, 29*(7), 90.

Winston, B. (2003). Extending Patterson's servant leadership model: Explaining how leaders and followers interact in a circular model. Regent University Servant Leadership Roundtable Regent University.

Chapter 12
Servant Leadership for Building the Community of Responders

Everyone desires relationships and community. Most people want to belong to a cohesive, like-minded group. It staves off loneliness. It promotes identity. These are natural and very human instincts.

—*Joshua Ferris*

Abstract This chapter brings together the philosophy of servant leadership as a way for building the community of responders. Emergency services leadership is accountable for this community. To be successful, the leader must become keenly aware of the vulnerabilities that threaten the responder. The profession is a noble one—one of a desire to serve strangers in their time of need. Though the responder is a highly trained and skilled operator, they still remain human. The community of responders has to be the place where servant leaders meet the needs of followers so that they can let down their armor and realize a long and rewarding career.

The notion of the "tough guy" mentality is nothing more than a myth sold through movies and books (May 1991). It is the idea that there are some humans that are so brave, so macho, that nothing can get to them. Within our society, we want to believe that our heroes are superheroes; that they are indestructible. However, it is just not the case, it is a legend, and something of pure fiction (May 1991). For years, people have assumed that those who perform emergency services work are not affected by the experiences. It is possible that because we put these individuals on pedestals, it makes it difficult for us to see our hero's as vulnerable. It pains us to admit that granite can crack.

Today, what used to be the dirty little secret within the profession is out there for everyone to see. Psychological research, some of which is sponsored by government agencies, has shined a light on the fact that responders are taking their own lives and burning out in record numbers (Henderson et al. 2016; Meyer et al. 2012; Sweeney 2012). Today, suicides have surpassed line-of-duty deaths among professional emergency services responders. It all comes down to the fact that you cannot remove an individual's humanity.

© Springer Nature Switzerland AG 2019

E. J. Russell, *In Command of Guardians: Executive Servant Leadership for the Community of Responders*, https://doi.org/10.1007/978-3-030-12493-9_12

Leaders within the emergency services have to commit to changing these statistics. It does not work when they intervene after the fact; it is too late at that point. Nobody wants to see responders suffer. The way forward is to serve this community of responders, strengthening their resolve before the traumas, and serving their needs within that community so they can grow. It is about their posttraumatic growth and it is about getting them to a point where they can transcend the long-term effects of trauma and instead grow as people from the experiences (Paton 2005). It starts by serving the community, for as Greenleaf (1977/2002) stated,

> All that is needed to rebuild community as a viable life form for large numbers of people is for enough servant-leaders to show the way, not by mass movements, but by each servant-leader demonstrating his or her unlimited liability for a quite specific community-related group (p. 53).

Veteran responders that have done the work are aware that evil exists. They have had to face it and they are all too aware of the things that go bump in the night. They know how fragile life is. Responders that have been around for a while understand that the only things that truly matter are people and relationships. The community of responders has to capture that in order to change the psychological impacts that go along with this type of work. It all has to be about the people.

The community of responders has to be a place where individuals can let down their armor and shelf their public façade. Yes, society asked them to be a certain way, to be sure of their abilities, to present themselves as capable and strong. But in private and off-duty, responders cannot keep up that persona (Thurnall-Read and Parker 2008). Individuals cannot remain "on point".

The Leadership Philosophy for the Emergency Services

Of the dozens of leadership theories and philosophies, servant leadership philosophy is the only approach that has been empirically tested within the profession. No other leadership theory or philosophy has ever been so researched within the emergency services career field. The characteristics and constructs that form the emergency responder's desire to serve have been found to match those of the servant leader (Spears 2010; Patterson 2003; Reed 2015; Russell 2014a, b; Russell et al. 2015). The philosophy mimics what it means to be an emergency services responder (Russell 2014a).

The philosophy of servant leadership has the power to change the emergency services. Emergency service leaders that embody the philosophy simply have to use what is inside of them already, harnessing their desire to serve, and then bring it to life in their leadership. For the emergency services leader, the philosophy is innate. Becoming a servant leader within the emergency services is a simple, natural progression. The process involves the leader allowing his or her personally held virtues to show outwardly to others. It is all about putting people first and serving their needs so they can serve the needs of others.

Addressing Skepticism

Here is a little truth about the philosophy of servant leadership that drives the skeptics crazy. Since Greenleaf (1970) penned his original essay almost 50 years ago, tens of thousands of servant leadership research studies, books, academic journal articles, and centers have been created. And in that time, not a single study has ever discovered that servant leadership does not work. The skepticism that surrounds the philosophy of servant leadership is nothing more than individual opinion. The skeptics often times get hung-up on the use of the word "servant" (Russell 2016). Skeptics against the philosophy misinterpret the word to mean "servitude" and then do their best to get others to dismiss it (Russell 2016).

To skeptics, servant leadership is seen as ice cream for breakfast. They miss the point that the servant leader is the one who benefits by serving his or her followers (Russell 2016). The servant leader realizes legitimate power, greater success, and greater wealth (Russell 2016; Russell et al. 2017). Why does the skepticism matter? Because the skepticism and the naysaying keep organizations and followers from being able to reach full potential (Sendjaya 2015). The push back against the philosophy will always come from those who seemingly buy into the "Great Man" theory which declares that leaders are born, not made. For those who study leadership, that belief is a fallacy that has been disproven. Leaders are cultivated over time through experience and learning (Russell 2014b).

For the leader that is looking for a successful pathway that benefits the organization, the people, and self, they need to take a close look at servant leadership (Turner 2000; Vinod and Sudhakar 2011). However, because of the skepticism that some have for the philosophy, the leader needs to be able to make the case and persuade others (Keith 2008; Thurnall-Read and Parker 2008).

A Community of Teamwork

The emergency services profession is about teamwork. The nature and demand of the career field warrants that functional and competent teams are vital in order to carryout operations (Salka and Neville 2004). The same goes for the community of responders away from the emergency scene. They need to be a team, they have to work together and support one another. As an emergency services leader, one is responsible for building and cultivating these teams.

Even though rank and responsibility change due to promotion, one is still a member of the team (Barker 2017; Sargent 2006). And although leaders are accountable for the overall organization, they are still an emergency responder and a part of this community; they just have to play a different role (Baker 2011; Smeby 2005).

A servant leader is one who builds great teams (Autry 2001). They understand the strength and positive outcomes that come from competent and capable teams (Autry 2001). The servant leader is keenly aware of the needs of individuals that come together to form those teams (Greenleaf 1977/2002). And because they are aware and they desire to serve, they ensure that the needs of the team members are met. In doing so, teams have what they need for success.

Community as a Place for Understanding

The community of responders has to be a place of understanding. It has to be a place of safety and acceptance. Responders have to feel at home in their community. To be in command of guardians, means that you are the guardian of the community. You are responsible for how it is shaped. You are responsible for openness, dialogue, and healing. As the guardian of the community, you have to protect it from the bureaucracy that is responsible for so many of the responder's problems (Kirschman 2004, 2006; Rhodes 2006).

A community of understanding begins with recognizing the human in people. It starts with admitting that the emergency responder has a life outside of work, and that no matter how good they are, sometimes that life creeps in. Divorces, family troubles, and tragedies are all impossible to leave at the door of the station. They don't enter their shift as an unfeeling machine; the community needs to start accepting that. This goes for emergency response as well. There are certain scenarios and situations that will remind the responder of their own life, their own family. It will have an effect on them; it doesn't even have to be a traumatic call, just one that they may deeply relate to. The only way to know when this is happening is through open dialogue and an awareness of who your people are.

Part of building a community of understanding is recognizing that sometimes people will have a bad day. There are times responders will be going through personal things that will affect their behaviors and abilities. A community of understanding is one that knows it is only temporary. The community will be there to support the responder, to get them through whatever it is they are going through, so they can come back better, stronger, and more loyal.

A community of understanding needs to be a place where responders feel valued. It has to be a place where they are treated as professionals and where their knowledge, skills, and abilities are respected (Covey 1997; Russell et al. 2015). The servant leader, believing in empowerment and fostering relationships built on trust, has no problem gifting responsibility to followers. People feel valued when they are empowered; it gives them that sense of ownership. When you value somebody as a professional and you respect their abilities, you openly delegate to them. Empowering responders lets them know they are valued and that their leadership believes in them.

The community of understanding knows that being an emergency services responder is part of an individual's identity. When their actions are called into question, the leader has to realize that they are calling their identity into question. That does not mean that disciplinary and corrective actions are not called for. It just means that the leader has to be aware that when they are criticizing a responder's actions, it is received as a criticism of him or her as an individual. However, when a trust-built relationship exists and open dialogue is allowed within an organization, individuals realize that the criticism that they are receiving is for their betterment and not a personal attack.

Summary

People need to have a sense of belonging, a sense of community (Maslow 1943). Part of being an emergency responder is belonging to the community of responders. That community is their family. Outside of the organization the community of responders as a whole becomes their extended family. Responders experience this when they travel, when they meet fellow responders. Though they are strangers, they interact as if they are friends. No matter what country they are in, or what language is spoken, emergency responders find kinship and familiarity with other responders. It is a bond forged in their shared desire to serve and sacrifice.

You, being in command of guardians, are a part of a small group of the executive level leaders that has been gifted the responsibility to lead an emergency services organization. At times it can be a daunting task. As a leader, your success is measured by the success of your people. Your followers are the ones that have to do the work. Who they are and the way that they conduct themselves is a direct reflection on not just the organization, but also you, as a leader. The only way that you can truly lead these people is to love and serve them so that they can serve others.

When one decides to spend time learning about the philosophy of servant leadership, they soon realize its potential power for transforming organizations and people (Blanchard and Hodges 2003). We know based upon statistics that leadership within the emergency services has to be transformed. Executive level leaders must focus on their followers; they must let go of mundane bureaucratic management that is far too often sold as leadership. This situation keeps leaders from focusing on their ultimate responsibility: serving those who serve. Serving the community of responders needs to become one's motivation (Covey 1997).

The premise of this work was to gift to those who are in command of guardians a mirror that allows inward reflection of both self and the organization. The emergency services profession can only benefit by this leadership philosophy. It has the potential of transforming the career field and the lives of responders. Bringing this philosophy into the profession begins with you calling upon your desire to serve and using it to become a servant leader in command of guardians.

Case Study

You are an executive level leader in command of guardians responsible for a community of responders.

Case Study Questions

1. What does it mean to you to be in command of guardians?
2. What type of leader do you desire to be and why?
3. What type of leader does the community of responders need and why?

References

Autry, J. (2001). *The servant leader: How to build a creative team, develop great morale and improve the bottom-line performance.* New York: Crown Business.

Baker, T. (2011). *Effective police leadership: Moving beyond management.* Flushing: Looseleaf Law Publications.

Barker, K. C. (2017). *Servant Leadership and Humility in Police Promotional Practices* (Doctoral dissertation, Walden University).

Blanchard, K., & Hodges, P. (2003). *The servant leader: Transforming your heart, head, hands, & habits.* Nashville: Thompson Nelson.

Covey, S. (1997). *First things first every day.* New York: Fireside.

Greenleaf, R. (1970). *The servant as a leader.* Indianapolis: Greenleaf Center.

Greenleaf, R. (1977/2002). *Servant-leadership: A journey into the nature of legitimate power and greatness.* Mahwah: Paulist Press.

Henderson, S. N., Van Hasselt, V. B., LeDuc, T. J., & Couwels, J. (2016). Firefighter suicide: Understanding cultural challenges for mental health professionals. *Professional Psychology: Research and Practice, 47*(3), 224–230.

Keith, K. (2008). *The case for servant leadership.* Westfield: Greenleaf Center for Servant Leadership.

Kirschman, E. (2004). *I love a firefighter: What every family needs to know.* New York: Guilford Press.

Kirschman, E. (2006). *I love a cop: What every family needs to know.* New York: Guilford Press.

Maslow, A. H. (1943). A theory of human motivation. *Psychological Review, 50,* 370–396.

May, R. (1991). *The cry for myth.* New York: W.W. Norton.

Meyer, E. C., Zimering, R., Daly, E., Knight, J., Kamholz, B. W., & Gulliver, S. B. (2012). Predictors of posttraumatic stress disorder and other psychological symptoms in trauma-exposed firefighters. *Psychological Services, 9*(1), 1–15.

Paton, D. (2005). Posttraumatic growth in protective services professional: Individual, cognitive and organizational influences. *Traumatology, 11,* 335–346.

Patterson, K. (2003). Servant leadership: A theoretical model (Doctoral Dissertation). Available from ProQuest Dissertation and Theses Database. (UMI No. 3082719).

Reed, L. (2015). Servant leadership, followership, and organizational citizenship behaviors in 9-1-1 emergency communications centers: Implications of a national study. *Servant Leadership: Theory and Practice, 2*(1), 71–94.

Rhodes, D. (2006). Katrina: "Brotherhood vs. bureaucracy". *Fire Engineering, 159*(5), 71.

Russell, E. (2014a). *The desire to serve: Servant leadership for fire and emergency services.* Westfield: Robert K. Greenleaf Center for Servant Leadership.

Russell, E. (2014b). Servant leadership theory and the emergency services learner. *Journal of Instructional Research, 3*(1), 64–72.

Russell, E. (2016). Servant leadership's cycle of benefit. *Servant Leadership: Theory & Practice, 3*(1), 52–68.

Russell, E., Broomé, R., & Prince, R. (2015). Discovering the servant in fire and emergency services leaders. *Servant Leadership: Theory & Practice, 2*(2), 57–75.

Russell, E., Maxfield, J., & Russell, J. (2017). Discovering the self-interest of servant leadership. *Servant Leadership: Theory & Practice, 4*(1), 75–97.

Salka, J., & Neville, B. (2004). *First in, last out: Leadership lessons from the New York Fire Department.* New York: Penguin.

Sargent, C. (2006). *From buddy to boss: Effective fire service leadership.* Tulsa: PennWell.

Sendjaya, S. (2015). *Personal and organizational excellence through servant leadership: Learning to serve, serving to lead, leading to transform.* New York: Springer International.

Smeby, C. (2005). *Fire and emergency services administration: Management and leadership practices.* Sudbury: Jones and Bartlett.

Spears, L. (2010). Servant leadership and Robert K. Greenleaf's Legacy. In K. Patterson & D. van Dierendonck (Eds.), *Servant leadership: Developments in theory and research*. New York: Palgrave Macmillan.

Sweeney, P. (2012). When serving becomes surviving: PTSD and suicide in the fire service. Retrieved from http://sweeneyalliance.org/.

Thurnall-Read, T., & Parker, A. (2008). Men, masculinities and firefighting: Occupational identity, shop-floor culture and organizational change. *Emotion, Space and Society, 1*(2), 127–134.

Turner, W. (2000). *The learning of love: A journey toward servant leadership*. Macon: Smyth and Helwys.

Vinod, S., & Sudhakar, B. (2011). Servant leadership: A unique art of leadership! Interdisciplinary. *Journal of Contemporary Research in Business, 2*(11), 456–467.

CPSIA information can be obtained
at www.ICGtesting.com
Printed in the USA
LVHW080833300321
682940LV00003B/95